Lecture Notes in Computer Science 1732

Edited by G. Goos, J. Hartmanis and J. van Leeuwen

Springer

Berlin
Heidelberg
New York
Barcelona
Hong Kong
London
Milan
Paris
Singapore
Tokyo

Satoshi Matsuoka   Rodney R. Oldehoeft
Marydell Tholburn (Eds.)

Computing
in Object-Oriented
Parallel Environments

Third International Symposium, ISCOPE 99
San Francisco, CA, USA, December 8-10, 1999
Proceedings

 Springer

Series Editors

Gerhard Goos, Karlsruhe University, Germany
Juris Hartmanis, Cornell University, NY, USA
Jan van Leeuwen, Utrecht University, The Netherlands

Volume Editors

Satoshi Matsuoka
Department of Mathematical and Computing Sciences
Tokyo Institute of Technology
2-12-1 Oh-okayama, Meguro-ku, Tokyo 152, Japan
E-mail:matsu@is.titech.ac.jp

Rodney R. Oldehoeft
ACL, MS B287, Los Alamos National Laboratory
Los Alamos, NM 87545, USA
E-mail: rro@lanl.gov

Marydell Tholburn
CIC-8, MS B272, Los Alamos National Laboratory
Los Alamos, NM 87545, USA
E-mail: marydell@lanl.gov

Cataloging-in-Publication data applied for

Die Deutsche Bibliothek - CIP-Einheitsaufnahme

Computing in object oriented parallel environments : third international
symposium ; proceedings / ISCOPE 99, San Francisco, CA, USA, December 8 - 10,
1999. Satoshi Matsuoka ... (ed.). - Berlin ; Heidelberg ; New York ; Barcelona
; Hong Kong ; London ; Milan ; Paris ; Singapore ; Tokyo : Springer, 1999
(Lecture notes in computer science ; Vol. 1732)
ISBN 3-540-66818-7

CR Subject Classification (1998): D, G.1-2, F.3, F.2

ISSN 0302-9743
ISBN 3-540-66818-7 Springer-Verlag Berlin Heidelberg New York

© Springer-Verlag Berlin Heidelberg 1999
Printed in Germany

Typesetting: Camera-ready by author
SPIN: 10704054   06/3142 – 5 4 3 2 1 0   Printed on acid-free paper

Preface

This volume contains the proceedings of the International Symposium on Computing in Object-Oriented Parallel Environments (ISCOPE 99), held in San Francisco, California, USA on December 8–10, 1999. ISCOPE is in its third year,[1] and continues to grow both in attendance and in the diversity of the subjects covered. The original ISCOPE meetings and the predecessor conferences focused more narrowly on scientific computing in the high-performance arena. ISCOPE 98 retained this emphasis, but broadened to include discrete-event simulation, mobile computing, and web-based metacomputing. ISCOPE 99 continues this trend.

The ISCOPE 99 program committee received 41 submissions, and accepted 14 (34%) as regular papers, based on their excellent content, maturity of development, and likelihood for widespread interest. In addition, the program committee selected six submissions as short papers. These papers were deemed to represent important work of a more specialized nature or to describe projects that are still in development.

The 20 papers are divided into seven technical categories:

Compilers and Optimization Techniques
New Application Areas
Components and Metacomputing
Numerical Frameworks
Generic Programming and Skeletons
Application-Specific Frameworks
Runtime Systems and Techniques

This collection of 20 papers represents today's state of the art in applying object-oriented methods to parallel computing. ISCOPE 99 is truly international in scope, with its 52 contributing authors representing 21 research institutions in 8 countries. The ISCOPE 99 organizers are confident that the reader will share their excitement about this dynamic and important area of computer science and applications research.

At the end of this volume, the author contacts section details the affiliations, postal addresses, and email addresses of all the proceedings authors.

October 1999

Satoshi Matsuoka
Rodney R. Oldehoeft
Marydell Tholburn

---

[1] The ISCOPE 97 and ISCOPE 98 proceedings are available from Springer as LNCS Volumes 1343 and 1505, respectively.

Steering Committee

Denis Caromel, University of Nice–INRIA Sophia Antipolis
Dennis Gannon, Indiana University
Yutaka Ishikawa, Real World Computing Partnership
Satoshi Matsuoka, Tokyo Institute of Technology
Jörg Nolte, German National Research Center for Information Technology
John Reynders, Los Alamos National Laboratory

Organizing Chairs

John Reynders, Los Alamos National Laboratory, General chair
Satoshi Matsuoka, Tokyo Institute of Technology, Program Chair
Yutaka Ishikawa, Real World Computing Partnership, Posters
Rodney R. Oldehoeft, Los Alamos National Laboratory, Proceedings
Marydell Tholburn, Los Alamos National Laboratory,
                    Local Arrangements/Publicity
Geoffrey Fox, Syracuse University, Java Grande Activities

Program Committee

Denis Caromel, University of Nice INRIA–Sophia Antipolis, France
Sid Chatterjee, University of North Carolina, USA
Andrew Chien, University of California–San Diego, USA
Jack Dongarra, University of Tennessee/Oak Ridge National Lab, USA
Geoffrey Fox, Syracuse University, USA
Dennis Gannon, Indiana University, USA
Rachid Guerraoui, University of Lausanne EPFL, France
Scott Haney, Los Alamos National Lab, USA
Yutaka Ishikawa, Real-World Computing Partnership, Japan
Jean-Marc Jezequel, IRISA/CNRS, France
L. V. Kale, University of Illinois–UC, USA
Carl Kesselman, University of Southern California/ISI, USA
Doug Lea, State University of New York–Oswego, USA
Charles Norton, NASA JPL, USA
Dan Quinlan, Lawrence Livermore National Lab, USA
Roldan Pozo, National Institute of Science and Technology, USA
Martin Rinard, Massachusetts Institute of Technology, USA
Mitsuhisa Sato, Real-World Computing Partnership, Japan
David Snelling, Fujitsu European Center for Information Technology, UK
Guy L. Steele, Jr., Sun Microsystems Labs, USA
Kenjiro Taura, University of Tokyo, Japan
Andrew Wendelborn, University of Adelaide, Australia
Katherine Yelick, University of California–Berkeley, USA

Table of Contents

Complex Numbers for Java

Michael Philippsen and Edwin Günthner

University of Karlsruhe, Germany

Abstract. Efficient and elegant complex numbers are one of the precon-
ditions for the use of Java in scientific computing. This paper introduces a
preprocessor and its translation rules that map a new basic type `complex`
and its operations to pure Java. For the mapping is insufficient to just
replace one `complex`-variable with two `double`-variables.
Compared to code that uses `Complex` objects and method invocations to
express arithmetic operations the new basic type increases readability
and it is also executed faster. On average, the versions of our benchmark
programs that use the basic type outperform the class-based versions by
a factor of 2 up to 21 (depending on the JVM used).

1 Introduction

In regular Java there is just one reasonable way to use complex numbers, namely
to write a class `Complex` containing two values of type `double`. Arithmetic operations have to be expressed by method invocations as shown in the following
code fragment. The alternative, to manually use two `double`-variables where a
complex number is needed, is too error-prone and too cumbersome to be accept-
able.

```
Complex a = new Complex(5,2);
Complex b = a.plus(a);
```

Class-based complex numbers have three disadvantages: Once written without
operator overloading, arithmetic operations are hard to read and maintain. Second, since Java does not support so-called value classes, object creation is slower
and objects need more memory than variables of a basic type. Arithmetic operations based on classes are therefore much slower than arithmetic on built-in
types. Even worse, method-based arithmetic causes frequent creation of temporary objects to return values. To return temporary arithmetic results with
basic types, no such object creation is needed. The third disadvantage is that
class-based complex numbers do not seamlessly blend with basic types and their
relationships. For example, an assignment of a `double`-value to a `Complex`-object
will not cause an automatic type cast – although such a cast would be expected
for a genuine basic type `complex`. Additionally, there is no natural way to express
complex literals; instead a constructor call is needed.

The fraction of people using Java for scientific computing is quite small, so
it is unlikely that the Java Virtual Machine (JVM) or the Java bytecode will

S. Matsuoka et al. (Eds.): ISCOPE'99, LNCS 1732, pp. 1–12, 1999.

be extended to support a basic type `complex` – although this might be the best solution from a technical point of view. It is also hard to tell whether Java will ever be extended to support operator overloading and value classes; and if so, whether there will be efficient implementations. But even given such features our work would remain important because, first, the same level of seamlessness cannot be achieved, see the above type cast problem. And second, our work can still be used to rate the efficiency of implementations of the general features.

The next section discusses the related work. Section 3 gives an overview of the preprocessor/compiler. The basic ideas of the translation are presented in Sect. 4. Section 5 shows the quantitative results.

2   Related Work

With support from Sun Microsystems, the Java Grande Forum [1,2] strives to improve the suitability of Java for scientific computing. The challenge is to identify and bundle the needs of this small user group in such a way that they can be respected in the continuing evolvement of Java although that is driven by the main stream.

The Java Grande Forum is working on a reference implementation of a class `Complex` that can be used to express arithmetic on complex numbers [3,4]. Special attention is paid to problems of numerical stability. IBM is extending their Java-to-native compiler to recognize the use of this class [5]. By understanding the semantics of the `Complex` class, the compiler can optimize away method invocations and avoidable temporary objects. Hence – at least on some IBM machines – high performance can be achieved even when using class-based complex numbers. However, the other disadvantages mentioned above still hold, i.e. there is no operator overloading and `Complex` objects lack a seamless integration into the basic type system.

There are considerations to add value classes to the official Java language [6,7]. But although there is no proper specification and no implementation yet, the Borneo project [8] is at least in a stage of planning. Since there are already object-oriented languages that support value classes, e.g. Sather [9], the basic technical questions of compiling value classes to native code can be regarded as solved. However, it is still unclear whether and how value classes can efficiently be added to Java by a transformation that expresses value classes with original language elements. In particular, it remains to be seen whether value classes will require a change of the bytecode format.

A primitive type `complex` is currently being added to C/C++ [10]. It is unclear whether a separate imaginary primitive type should be added to the language, as Kahan proposes in [11].

3   *Cj* at a Glance

In our experience, any extension of Java will only be accepted if there is a transformation back to pure Java. But in general better efficiency can be achieved by using optimization techniques during bytecode generation. Our compiler *cj*, which is an extension of *gj* [12], therefore supports two different output formats, namely Java bytecode and Java source code. In this paper, we focus on the latter and mention optimizations only briefly.

Cj extends the set of basic types by the type `complex`. A value of type `complex` represents a pair of two double precision floating point numbers. All the common operations for basic types are defined for `complex` in a straightforward way. The real and imaginary part of a `complex` can be accessed through the member fields `real` and `imag`. Note, that the names of these fields are no new keywords. Since the basic type `complex` is a supertype of `double`, a double-value will be implicitly casted where a `complex` is expected. A second new keyword, I, is introduced to represent the imaginary unit and to express constant expressions of type `complex`.

4   Recursive Transformation Rules for `complex`

For simplicity, we call any expression that uses `complex` values a `complex` expression. The transformation of `complex` expressions to expressions that only use `double` causes several problems.

• **Transformation locality.** If a `complex` expression is used where only an *expression* is allowed it must not be mapped to a sequence of statements.

```
while (u == v && x == (y = foo(z))) {...}
```

For example, for transforming the `complex` condition of this `while`-loop into its real and imaginary parts it is necessary to introduce several temporary variables whose values have to be calculated within the body of the loop. Therefore, the loop has to be reconstructed completely.

In general, to replace `complex` expressions by three-address statements, one needs non-local transformations that reconstruct surrounding statements as well, although local transformation rules that replace expressions by other expressions (not statements) were simpler to implement in a compiler and it would be easier to reason about their correctness.

• **Semantics.** To achieve platform independence, Java requires a specific evaluation order for expressions (from left to right). Any transformation of `complex` arithmetic must implement this evaluation order using `double`-arithmetic.

To preserve these semantics for `complex` expressions, it is not correct to fully evaluate the real part before evaluating the imaginary part. Instead, the transformation has to achieve that a side effect is only visible on the right hand side of its occurrence, but both for the real and imaginary part. Similarly in case of an exception, only those side effects are to become visible that occur on the left side of the exception. Additionally, by separating the real part from the imaginary part, it is also unclear how to treat method invocations (`foo(z)` in the above example). Shall `foo` be called two times? Is it even necessary to create two versions of `foo`?

4.1 Sequence Methods

To avoid both types of problems we introduce what we call *sequence methods* as a central idea of *cj*. Each `complex` expression is transformed into a sequence of expressions. These new expressions are then combined as arguments of a sequence method. The return value of a sequence method is ignored. This technique enables us to keep the nature of an expression and allows our transformation to be local. A sequence method has an empty body; all operations happen while evaluating the arguments of the method invocation. The arguments are evaluated in typical Java ordering from left to right.[1] In case of nested expressions, the arguments of a sequence method invocation are again invocations of sequence methods. By using this concept we are able to evaluate both parts of each node of a `complex` expression tree at a time. Moreover, the evaluation order (in terms of visibility of side effects and exceptions) is guaranteed to be correct.

When *cj* is used as preprocessor and Java code is produced, the method invocations of the sequence methods – which are declared `final` in the surrounding class – are not removed. However, they may be inlined by a Just-in-time (JIT) compiler. When *cj* generates bytecode, the compiler directly removes the method invocations – only the evaluations of the arguments remain.

An Example of Sequence Methods. Let us first consider the right hand side of the `complex` assignment z = x + y. To avoid any illegal side effects we use temporary variables to store all operands. The following code fragment shows the (yet unoptimized) result of the transformation of the right hand side.

```
seq(seq(tmp1_real = x_real, tmp1_imag = x_imag),
    seq(tmp2_real = y_real, tmp2_imag = y_imag),
    tmp3_real = tmp1_real+tmp2_real, tmp3_imag = tmp1_imag+tmp2_imag)
```

In this example, six `double`-variables would have to be declared in the surrounding block (not shown in the code). When evaluating this new expression, Java will start with the inner calls of sequence methods (from left to right). Thus, both parts of x and y are stored in temporary variables. The subsequent call of the enclosing sequence method performs the addition (in the third and fourth argument). A subsequent basic block optimization detects the copy propagation and eliminates passive code. So we only need a minimal number of temporary variables and copy operations. In the example, just two temporary variables and one sequence method remain.

```
seq(tmp3_real = x_real+y_real, tmp3_imag = x_imag+y_imag)
```

Now look at the assignment to z and the two required elementary assignments.

```
seq(seq(tmp3_real = x_real+y_real, tmp3_imag = x_imag+y_imag),
    z_real = tmp3_real, z_imag = tmp3_imag)
```

---

[1] Exceptions are not thrown within a sequence method but within the invocation context. Hence, it is unnecessary to declare any exceptions in the signature of sequence methods.

In this case the basic block optimization also reduces the number of temporary variables and prevents the declaration of a sequence method. Thus, the resulting Java code does not need any temporary variables; only a single sequence method needs to be declared in the enclosing class.[2]

```
seq(z_real = x_real + y_real, z_imag = x_imag + y_imag)
```

When we directly construct bytecode, there is no need for the sequence method. Instead, only the arguments are evaluated.

4.2 Basic Transformation Rules in Detail

In the next sections we consider an expression E that consists of subexpressions e_1 through e_n. The rewriting rule $eval[E]$ describes (on the right hand side of the $\mapsto$-symbol) the recursive transformation into pure Java that applies $eval[e_i]$ to each subexpression. In most cases complex expressions are mapped to calls of sequence methods whose results are ignored. Sometimes it is necessary to access the real or imaginary part of a complex expression. For this purpose there are $evalR$ and $evalI$. Both cause the same effect as $eval$ but are mapped to special sequence methods (seqREAL or seqIMAG) that return the real or the imaginary part of the complex expression. If $evalR$ or $evalI$ are applied to an array of complex-values the corresponding sequence methods will return an array of double-values. Expressions that are not complex remain unchanged when treated by $eval$, $evalR$, or $evalI$. The =-symbol refers to Java's assignment operator. In contrast, we use $\equiv$ to define an identifier (left hand side of $\equiv$) that has to be expanded textually by the expression on the right hand side.

To process the left hand side of assignments we use another rewriting rule: $access[E]$ does not return a value but instead returns the shortest access path to a subexpression, requiring at most one pointer dereferencing.

From the above example, it is obvious that a lot of temporary variables are added to the block that encloses the translated expressions. Most of these temporary variables are removed later by optimizations.[3] The following transformation rules do not show the declaration of temporary variables explicitly. However, they can easily be identified by means of the naming convention: if e is a complex expression, the identifiers e_{real} and e_{imag} denote the two corresponding temporary variables of type double. The use of any other temporary variables is explained in the text. Arrays of complex are discussed in Sect. 4.3; method invocations are described in Sect. 4.4. The rules for unary operations, constant values, and literals are trivial and will be skipped. Details can be obtained from [13].

- **Plain identifier:** The transformation rule for $E \equiv c$ is:
$$eval[c] \mapsto seq(E_{real} = c_{real}, E_{imag} = c_{imag})$$

---

[2] Since user defined types may appear in the signature of sequence methods it is impossible to predefine a collection of sequence methods in a helper class.

[3] In case of static code or the initialization of instance variables the remaining temporary variables are neither static nor instance variables: they can be converted to local variables by enclosing them with static or dynamic blocks.

Both components of the `complex` variable c are stored to temporary variables that represent the result of the expression E. If c is used as left hand side of an assignment, it is sufficient to use the mangeled names.

- **Selection:** The transformation rule for $E \equiv F.e$ is:

$$eval[F.e] = seq(tmp = eval[F], E_{real} = tmp.e_{real}, E_{imag} = tmp.e_{imag})$$

F is evaluated once and stored in a temporary variable tmp and then tmp is used to access the two components.

If $F.e$ is used as the left hand side of an assignment, F is evaluated to a temporary variable that is used for further transformations of the right hand side:

$$access[F.e] \mapsto tmp = eval[F]$$
$$\wedge\ E_{real}^{\downarrow} \equiv tmp.e_{real}, E_{imag}^{\downarrow} \equiv tmp.e_{imag}$$

It is important to note that the transformation rule for assignments (see below) demands that the code on the right hand side of the $\mapsto$-symbol is inserted at the position where $access[F.e]$ is evaluated. Secondly, the identifiers $E_{real}^{\downarrow}$ and $E_{imag}^{\downarrow}$ have to be replaced textually with the code following the $\equiv$-symbol. (The $\downarrow$-notation and the textual replacement are supposed to help understanding by clearly separating the issues of the access path evaluation from the core assignment.)

- **Assignment:** The transformation rule for $E \equiv e_1 = e_2$ is:

$$eval[e_1 = e_2] \mapsto seq(access[e_1], eval[e_2], E_{real} = e_{1real}^{\downarrow} = e_{2real}, E_{imag} = e_{1imag}^{\downarrow} = e_{2imag})$$

First the access to e_1 is processed. Then the right hand side of the assignment is evaluated. The last two steps perform the assignment of both parts of the `complex` expression. Since the assignment itself is an expression it is necessary to initialize additional temporary variables that belong to E. Occurrences of $e^{\downarrow}$ are inserted textually according to $access$.

Therefore, the transformation creates the following code for `X.Y.z = x` (after removing temporary variables and redundant calls of sequence methods):

```
seq(tmp = X.Y, tmp.z_real = x_real, tmp.z_imag = x_imag)
```

The temporary variable `tmp` is only necessary if `X.Y.` may cause side effects.

- **Combination of assignment and operation:** The transformation rule for $E \equiv e_1 \diamond = e_2$, where $\diamond \in \{+, -, *, /\}$, is:

$$eval[e_1 \diamond = e_2] = seq(access[e1], e_1^{\downarrow} = eval[e_1^{\downarrow} \diamond e_2])$$

This strategy is essential to avoid repetition of side effects while evaluating e_1.

- **Comparison:** The transformation rule for $E \equiv e_1 == e_2$ is:

$$eval[e_1 == e_2] \mapsto seq_{value}(eval[e_1], eval[e_2], e_{1real} == e_{2real}\ \&\&\ e_{1imag} == e_{2imag})$$

In contrast to the sequence methods used before, this one is not returning a dummy value. Instead seq_{value} returns the value of its last argument. The result of the whole expression is a logical AND of the two comparisons. Inequality tests can be expressed in the same way, we just have to use `!=` and `||` instead of `==` and `&&`. This special kind of sequence method can also be removed while generating bytecode.

- **Addition and subtraction:** The transformation rule for $E \equiv e_1 \diamond e_2$, where $\diamond \in \{+, -\}$, is:

$$eval[e_1 \diamond e_2] \mapsto seq(eval[e_1], eval[e_2], E_{real} = e_{1real} \diamond e_{2real}, E_{imag} = e_{1imag} \diamond e_{2imag})$$

- **Multiplication:** The transformation rule for $E \equiv e_1 * e_2$ is:
$$eval[e_1 * e_2] \mapsto seq(eval[e_1], eval[e_2], E_{real} = e_{1real} * e_{2real} + e_{1imag} * e_{2imag},$$
$$E_{imag} = e_{1real} * e_{2imag} - e_{1imag} * e_{2real})$$

- **Division:** The rule for division is structurally identical to the rule for multiplication but the expressions are considerably more complicated. cj offers two versions to divide `complex` expressions: a standard implementation and a slower but numerically more stable version. The second alternative is based on the reference implementation [3]. For brevity, none of the versions is shown.

- **Type cast:** Because `complex` is defined as a supertype of `double`, implicit type casts are inserted where necessary. Furthermore, it is appropriate to remove explicit type casts to `complex` if the expression to be casted is already of type `complex`. The case $(E \equiv (complex)\ e)$ can be handled with the following rule:
$$eval[(complex)\ e] \mapsto seq(eval[e], E_{real} = e, E_{imag} = 0)$$

- **String concatenation:** Since string concatenation is not considered as time-critical cj creates an object of type `Complex` and invokes the corresponding method `toString`. An additional benefit is that the output format can be changed without modifying the compiler. The transformation rule is:
$$eval[str + e] \mapsto str + (new\ Complex(evalR[e], e_{imag}).toString())$$
$EvalR[e]$ evaluates e and returns its real part. Furthermore $evalR$ declares a temporary variable e_{imag} and initializes it with the imaginary part of e. This asymmetry is necessary to ensure that e is evaluated exactly once. For brevity, we skip similar rules for $e + str$ and the $+ =$-operation.

4.3   Transformation Rules for Arrays

Although it is obvious that a variable of type `complex` must be mapped to a pair of two `double`-variables, there is no obvious solution for arrays of `complex`. There are two options: an array of `complex` can either be replaced by two `double`-arrays or by one `double`-array of twice the size. For the latter, our performance measurements indicated that in general it is faster to store pairs of `double`-values in adjacent index positions than to store all real parts *en bloc* before storing all the imaginary parts.

Since the 2-array solution is neither faster nor slower than the 1-array solution as shown in Fig. 1 and since use of the 2-array solution eases the implementation of cj we selected.

For various array sizes, we compared the speed of array creation, initialization and garbage collection (called init in Fig. 1) for the 1-array and the 2-array solution. The lines show how much faster (< 1) or slower (> 1) the 2-array solution is over the 1-array solution ($= 100\%$). Similar for read and write access to the array elements. All measurements have been repeated several times to be able to ignore clock resolution and to achieve a small variance.

It can be noticed that there is no clear advantage of either the 1-array solution of the 2-array solution. We got similar results on other platforms and with other JITs. The average of all measurements is within $[0.98; 1, 02]$ for both platforms. Read and write access is much more stable with Hotspot, however there is some

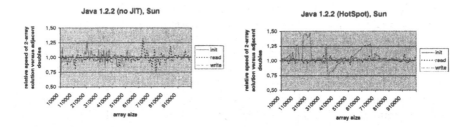

Fig. 1. Performance of two options to deal with arrays

peculiar behavior for initialization. Hotspot probably uses different mechanisms for arrays of different sizes.

• **Array creation and initialization:** Java offers different language elements to create arrays or to create and initialize arrays in one step. Let us first discuss the transformation rule for pure array creation:

$$eval[new\ complex[e_1]\dots[e_n]] \mapsto$$
$$seq(E_{real} = new\ double[e'_1 = eval[e_1]]\dots[e'_n = eval[e_n]],$$
$$E_{imag} = new\ double[e'_1]\dots[e'_n])$$

When calculating E_{real}, additional temporary variables e'_i are used to allow the reuse of size expressions in the imaginary part. Array initialization is done according to the following rule:

$$eval[new\ complex\ []\dots[]\{e_1,\dots,e_n\}] \mapsto$$
$$seq(E_{real} = new\ double[]\dots[]\{evalR[e_1],\dots,evalR[e_n]\},$$
$$E_{imag} = new\ double[]\dots[]\{e_{1imag},\dots,e_{n_{imag}}\})$$

EvalR, is able to handle inner array initialization by applying the same rule recursively to array initializations with a smaller number of dimensions.

• **Array access used as left hand side of an assignment:** Such expressions may be affected by side effects because the evaluation of index expressions could modify the array itself. To avoid this problem it is in general necessary to store a reference to the array in a temporary variable (more exactly: we store a reference to $F[e_1]\dots[e_{n-1}]$). The transformation rule for the general case $(n > 1)$ is:

$$access[F[e_1]\dots[e_n]] \mapsto tmp = eval[F[e_1]\dots[e_{n-1}]]$$
$$\wedge\ E^{\downarrow}_{real} \equiv tmp_{real}[e'_n = eval[e_n]],\ E^{\downarrow}_{imag} \equiv tmp_{imag}[e'_n])$$

Again we are using new temporary variables e'_i to make sure that index expressions are evaluated exactly once. The $\downarrow$-notation emphasizes that the given expressions are to be textually inserted on the right hand side of the assignment. If the given array is one-dimensional our rule can be simplified to: $tmp = eval[F]$.

• **Array access:** The transformation rule for $E \equiv F[e_1]\dots[e_n]$ is:

$$eval[F[e_1]\dots[e_n]] \mapsto seq(eval[F], E_{real} = F_{real}[e'_1 = eval[e_1]]\dots[e'_n = eval[e_n]],$$
$$E_{imag} = F_{imag}[e'_1]\dots[e'_n])$$

4.4 Transformation Rules for Method Calls

We discuss `complex` parameters and `complex` return values separately. Moreover, constructors must be treated differently.

- **Complex return value:** There are no means in the JVM instruction set to return two values from a method. An obvious work-around would be to create and return an object (or an array of two `doubles`) every time the method is called. In most cases, this object is only necessary to pass the result out of the method and can be disposed right afterwards. In contrast, *cj* creates a separate array of two `doubles` for each textual method call. This array is not defined in the enclosing block but at the beginning of the method that encloses the call. This strategy minimizes the number of temporary objects that have to be created, e.g. for a call inside a loop body. Instead of calling the original method *foo* we are calling a method $\widehat{foo}$ with a modified signature: we pass a reference to this temporary array as an additional argument. This temporary array is created once per call of the enclosing method and may be reused several times. So the transformation rule for $E \equiv foo()$ is (similar for methods with arguments):

$$eval[foo()] \mapsto seq(\widehat{foo}(tmp), E_{real} = tmp[0], E_{imag} = tmp[1])$$

Two details are important to ensure the correctness of this transformation for recursive calls and in multithreaded situations: First, the temporary array is local to the enclosing method and second, every textual occurrence of a call of *foo* causes the creation of a different temporary variable.

The return type of $\widehat{foo}$ is not `void`. Instead it returns a dummy value (`null`) so that it still can be used inside expressions.[4]

- **Complex argument:** We use the obvious approach by again modifying the signature of the method. Instead of passing one argument of type `complex` we hand over two `double`-values. It is important that not only the argument list of the method but also its name is changed. This is necessary to avoid collisions with existing methods that have the same argument types as the newly created one. The transformation rule can be formalized as (similar for methods expecting several arguments of type `complex`):

$$eval[bar(e)] \mapsto \widehat{bar}(evalR[e], e_{imag})$$

- **Constructor method:** Since the first statement in the body of a constructor needs to be a call of another constructor, under certain circumstances the techniques described above would cause illegal code. *Cj* solves this problem by generating an additional constructor that declares the required temporary variables in its signature. Due to space restrictions we have to refer to [13] for details.

5   Benchmarks

On a Pentium 100 with 64 MB of RAM and 512 KB of cache we have installed two operating systems: Linux 2.0.36 (Suse 6.0) and Windows NT Version 4 (service pack 4). We have studied several different Java virtual machines for our tests: a pre-release of SUN's JDK 1.2 for Linux, SUN's JDK 1.2.1 for Windows, a JDK from IBM, the JVM that is included in Microsoft's Internet Explorer 5, and the beta release of SUN's new JIT compiler HotSpot.

Our benchmarks fall into two groups: the group of kernel benchmarks measures array access, basic arithmetic on `complex`, and method invocations with

---

[4] Before returning, the elements of the newly added array argument are initialized.

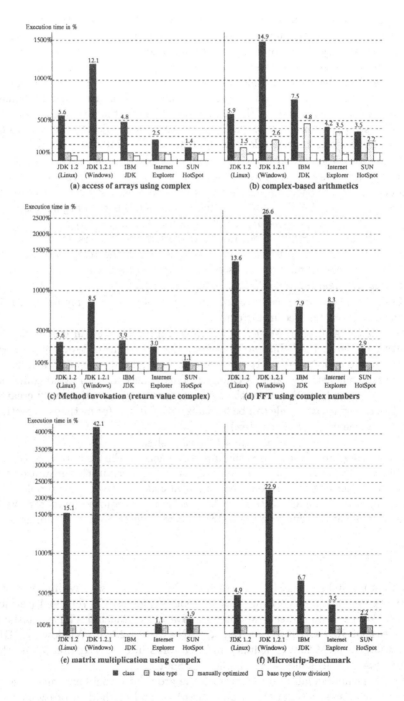

Fig. 2. Results of the benchmark programs

`complex` return values. The other group measures small applications: Microstrip calculations, `complex` matrix multiplication, and `complex` FFT. There are at least two versions of each program – one uses our basic type `complex` and the other uses a class to represent complex numbers.

5.1 Results

On average over all benchmarks, the programs using the basic type `complex` outperform the class-based versions *by a factor of 2 up to 21*, depending on the JVM used. We achieve the best factor with SUN's JDK 1.2 for Windows, which is the slowest JVM in our study. The smaller improvement factors are achieved with better JVMs (HotSpot and Internet Explorer) that incorporate certain optimization techniques, e.g., removal of redundant boundary checks, fast creation and handling of objects, and aggressive inlining of method bodies.

Figure 2 gives an overview over all benchmarks, labeled (a) to (f). In each of these six sub-figures there are five groups of bars, each group represents a different JVM. The most important item within a group is the black bar. This bar shows the relative execution time of the class-based version. The factor by which this version is slower than the basic type version (grey bar) is printed on top of the black bar. Some groups have more than two bars: here we did an additional transformation by hand, substituting each `complex` by two variables of type `double`. Those manually optimized programs (white bar) are just slightly faster than code generated by *cj*.

In sub-figures (a) to (c) the improvement is smaller than in the other figures. It is also apparent that better implementations of the JVM (Internet Explorer and HotSpot) are quite good in eliminating the overhead of object creation within the class-based solutions. But *cj* still performs better by 10% to 40%.

Benchmarks (a) and (c) focus on array access and method invocation. In contrast, programs (b) and (d-f) are predominantly calculating arithmetic expressions, where (d) and (e) are also showing some amount of array accesses. For arithmetic the techniques applied by *cj* (inlining of all method invocations and reducing the number of temporary variables) perform noticeably better than the class-based solution. Even on the better JVMs *cj* is 3 times faster. On slow JVMs *cj* achieves a factor of 8 or more. The main reason is that *cj* does a better inlining and can avoid temporary objects almost completely.

6 Conclusion

Complex numbers can be integrated seamlessly and efficiently into Java. Because of Java's strict evaluation order it is by far not enough to simply double the operations for their real and imaginary parts. Sequence methods enable a formalization of the necessary program transformations in a local context. Our technique for dealing with complex return values is efficient because it avoids the creation of many temporary objects. In comparison with their class-based counterparts, the benchmark programs that use the new primitive type perform better by a factor of 2 up to 21, on average, depending on the JVM used.

Acknowledgements

The Java Grande Forum and Siamak Hassanzadeh from Sun Microsystems supported us. Thanks to Martin Odersky for providing *gj*. Bernhard Haumacher and Lutz Prechelt gave valuable advice for improving the presentation.

References

1. Java Grande Forum. http://www.javagrande.org.
2. G. K. Thiruvathukal, F. Breg, R. Boisvert, J. Darcy, G. C. Fox, D. Gannon, S. Hassanzadeh, J. Moreira, M. Philippsen, R. Pozo, and M. Snir (editors). Java Grande Forum Report: Making Java work for high-end computing. In *Supercomputing'98*, Orlando, Florida, November 1998. panel handout.
3. Visual Numerics. Java grande complex reference. http://www.vni.com/corner/garage/grande/index.html, 1999.
4. IBM. Numerical intensive java. http://www.alphaWorks.ibm.com/tech/ninja/.
5. P. Wu, S. Midkiff, J. Moreira, and M. Gupta. Efficient support for complex numbers in Java. In *ACM 1999 Java Grande Conference*, pages 109–118, San Francisco, 1999.
6. J. Gosling. The evolution of numerical computing in Java. http://java.sun.com/people/jag/FP.html.
7. G. Steele. Growing a language. In *Proc. of OOPSLA'98*, October 1998. key note.
8. J. D. Darcy and W. Kahan. Borneo language. http://www.cs.berkeley.edu/~darcy/Borneo.
9. S. M. Omohundro and D. Stoutamire. The Sather 1.1 specification. Technical Report TR-96-012, ICSI, Berkeley, 1996.
10. C9x proposal. ftp://ftp.dmk.com/DMK/sc22wg14/c9x/complex/ and http://anubis.dkuug.dk/jtc1/sc22/wg14/.
11. William Kahan and J. W. Thomas. Augmenting a programming language with complex arithmetics. Technical Report No. 91/667, University of California at Berkeley, Department of Computer Science, December 1991.
12. G. Bracha, M. Odersky, D. Stoutamire, and P. Wadler. Making the future safe for the past: Adding genericity to the Java programming language. In *Proc. of OOPSLA'98*, October 1998. http://www.cis.unisa.edu.au/~pizza/gj/.
13. JavaParty. http://wwwipd.ira.uka.de/JavaParty/.

Automatic Generation of Executable Data Structures

Nobuhisa Fujinami

Sony Corporation, Japan

Abstract. The executable data structures method is a kind of run-time code generation. It reduces the traversal time of data structures by implementing the traversal operation as self-traversal code stored in each data structure. It is attractive but not easy to use. This paper shows that it can be viewed as run-time optimization of a group of objects and can automatically be generated from the source program in an object-oriented language. The encapsulation mechanism of object-oriented languages helps to analyze the program and to use the automatically generated code generators. Preliminary evaluation results show that the proposed method is better than conventional run-time code generation or manual source level optimization.

1   Introduction

Run-time code generation improves programs by generating machine code specific to values that are unknown until run time. Such values are called *run-time constants*. Examples are intermediate computation results and user inputs. Executing the generated code enough times will recoup the cost of run-time code generation.

Massalin's executable data structures (EDSs) method used in Synthesis Kernel [1,2] is a kind of run-time code generation. It reduces the traversal time of data structures by implementing the traversal operation as self-traversal code stored in each data structure. The code is specialized to both the traversal operation and the containing data structure. It is attractive but not easy to use, because it is implemented using hand-written templates in assembly language. Only well-trained programmers can use it.

This paper shows that EDSs can be viewed as an optimized implementation of a group of objects and proposes a method to generate them automatically from the source program in an object-oriented language. It assumes that the group consists of one representative object (container) and other objects (elements). The optimized program generates code for methods of the elements at run time. The code is optimized for both the element values and the caller method of the container. Partial modification of the group, e.g. adding a new element, is possible with a reasonable cost.

Straightforward application of the conventional run-time code generation framework does not lead to this optimization. Since it does not allow partial modification of the generated code, it can either:

S. Matsuoka et al. (Eds.): ISCOPE'99, LNCS 1732, pp. 13–24, 1999.

– optimize the container's method viewing all the elements and their relationships to be run-time constants, or
– optimize only the elements' methods.

The former makes it hard to modify the group, and the latter looses the opportunities of optimization.

The author has already proposed a system to automate run-time code generation in object-oriented languages [3,4]. The method proposed in this paper is an extension of the proposal. It is currently implemented as a preprocessor of C++ compilers. It focuses on instance variables whose values are invariant or quasi-invariant. For methods that use such values, it generates specialized code generators in C++[1]. It also inserts code fragments to invoke them and to invoke or invalidate the generated code. Whether the embedded instance variables are invariant and changes to the variables are automatically detected using the encapsulation mechanism of the object-oriented language[2]. Since run-time code generators are specialized for particular methods, code generation is light-weight. And since each instance has its own generated code for its methods, there can be multiple optimized versions for one method. Memory management for generated code is simple, because it can be left to the object construction/destruction mechanism of object-oriented languages.

The remainder of this paper is organized as follows. Section 2 overviews the idea of the EDSs method. Section 3 illustrates the optimization of a group of objects, and Sect. 4 describes a general method. Section 5 presents a preliminary evaluation of the method. Section 6 compares this paper with related work. Finally, Sect. 7 concludes the paper.

2   Executable Data Structures (EDSs)

The EDSs method reduces the traversal time of data structures. It is useful for data structures that are frequently traversed in a preferred way. If the data structures are nodes linked by pointers, the method stores node-specific traversal code along with the data in each node and makes the data structure *self-traversing*.

Figure 1 is an example of a task queue managed by a simple round-robin scheduler. Each element in the queue contains two short sequences of code: sw_out and sw_in. The sw_out saves the registers and branches to the next task's sw_in routine (in the next element in queue). The sw_in restores the new task's registers, installs the address of its own sw_out in the timer interrupt vector table, and resumes processing.

The sw_out and sw_in routines are optimized to the particular task and move only needed data. For example, if floating point operations have not been used in the task, floating point context is not saved/restored.

When adding a new element (task) to the queue, sw_out and sw_in routines specialized to the new element are generated using templates in assembly language, and the jump address in the sw_out routine in the previous element is set.

---

[1] When invoked, the code generators write machine instructions directly into memory.

[2] If the embedded values are changed, the code is invalidated and then regenerated.

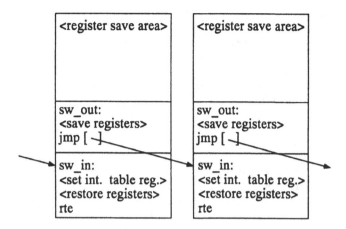

Fig. 1. Task queue

3 Optimizing a Group of Objects

In the rest of this paper, C++ notation is used for object-oriented terminology. For example, methods are called member functions, and instance variables are called data members. Assume also that all source code for implementing the object group is available.

The target of the EDSs method is a group of data structures frequently traversed in the similar way. If the group is implemented in an object-oriented language, the usual practice is to make one container object for the group. All operations on the group are invoked through the container. Various kinds of data structures, such as linked lists, binary trees, or hash tables can be implemented as this type of object groups.

If the conventional framework of run-time optimization is applied to the whole data structure represented by the objects, all the elements and their relationships are regarded to be run-time constants. In so doing, member functions of the elements used in the container's member functions can be inlined, when the machine code routines for the container's member functions are generated at run time.

The resulting code is efficient, but not flexible. The whole generated code becomes invalid each time when the relationships of the elements changes, when a new element is added, etc. Since the optimization result is a set of EDSs in the proposed method, the generated code allows partial modification. Adding a new element requires run-time code generation for the new element only.

For instance, Fig. 2 shows class definitions for a linked list in C++. Each element holds a string object. First, the list is empty. Member function **add** puts an element at the top of the list. Member function **search** tests if the parameter string is in the list or not.

Since the container **StringList** does not import any elements, it can only call member function **match** of its elements **StringLink** instantiated within it.

```
class String {
private:
  char s[MAXLEN];
  int len;
public:
  String(char const *p);
  int match(char const *p);
};

class StringLink {
public:
  StringLink *next;
  String s;
  StringLink(char const *p): s(p) {}
  int match(char const *p) { return s.match(p); }
};

class StringList {
private:
  StringLink *top;
public:
  StringList(): top(NULL) {}
  void add(char const *p);
  int search(char const *p);
private:
  int search(char const *p,StringLink *q);
};
```

Fig. 2. Class definitions for linked lists

Function match can be adapted to the container's member function search (see Fig. 3 for its definition). Function search can be implemented as distributed code fragments in the elements as shown in Fig. 4. The tail recursion of search is completely unrolled and the transition to the next element operation is implemented as a jump instruction to the code for the next element. The destination of the jump instruction is updated when the list is reconfigured.

Each of code fragments STRING1, STRING2, ··· is specialized to both its corresponding string and member function search. The if-statements in STRINGn is the result of specializing match with respect to the n-th string in the list. The rest is a specialized form of search.

Data members top_code and next_code in the code in Fig. 4 are newly added members to classes StringList and StringLink. They represent addresses of the code for the elements. Member functions that modify the linked list should update these members. When an instance of class StringList is initialized, the address of label NOSTRING in Fig. 4 should be assigned to member top_code. These assignments are parallel to the corresponding members top and next. Figure 5 shows member function add before and after the optimization.

```
int StringList::search(char const *p) { return search(p,top); }
int StringList::search(char const *p,StringLink *q) {
  if(q==NULL) return 0;
  if(q->match(p)) return 1;
  return search(p,q->next);
}
```

Fig. 3. Member function **search**

4 Automatic Optimization

This section describes a method for automating the optimization illustrated in
the previous section. It decides the applicability of the optimization, generates
run-time code generators, and modifies the program to use the code generators
and the generated code. The container and its elements are assumed to have
invariant data members.

4.1 Deciding Applicability

To use the EDSs method, some member functions of the container should repeatedly use its elements. This paper focuses on tail recursions as the iteration
constructs and on traversing links as a way to find next elements in iteration.
Non-tail recursions, such as full tree traversals, are not considered because they
require backward links of the generated code[3]. Loops are not considered either.
In most cases, loops can be rewritten into tail recursions by analyzing loop control variables.

The condition for finding iteration by tail recursions is as follows:

- There exists a container's member function, say r, that calls itself as tail
  recursions (there can be multiple calling sites).
- One of the formal parameters, say p, of r is a pointer to an element[4].
- Function r uses invariant data members of the object pointed to by p or
  invokes member functions[5] that use such data members.
- There exists a container's member function, say s, that calls r.

It is out of the scope of this optimization to invoke r directly from outside.
Container classes normally have member functions equivalent to s to encapsulate
internal implementation.

Though it is possible to deal with imported elements if the timing of impor-
tation is known, they are not considered for simplicity. All the elements used in
s and r should be instantiated in the member functions of the container. One
sufficient condition is to confirm all of the following:

---

[3] Traversing all the tree nodes must be rare cases, because tree structures are useful
   in reducing the number of nodes to be visited.

[4] Since reference types can be rewritten using pointer types, they are treated in the
   similar way.

[5] The function body will be inlined.

```
//---code in container---
StringList_search:
        <function prologue>
        p = this->top;                  // p <- address of first string
        return (*this->top_code)();     // call code for first string
        <function epilogue>
//---code in element STRING1 (string: "AB")---
STRING1: if(p->s.len!=2) goto NG1;      // string comparison
        if(p->s.s[0]!='A') goto NG1;
        if(p->s.s[1]!='B') goto NG1;
OK1:    return 1;
NG1:    tmp = p->next_code;
        p = p->next;                    // p <- address of next string
        goto *tmp;                      // jump to code for next string
//---code in element STRING2 (string: "C")---
STRING2: if(p->s.len!=1) goto NG2;      // string comparison
        if(p->s.s[0]!='C') goto NG2;
OK2:    return 1;
NG2:    tmp = p->next_code;
        p = p->next;                    // p <- address of next string
        goto *tmp;                      // jump to code for next string
//---termination code---
NOSTRING: return 0;
```

Fig. 4. Optimized member function **search**. Pseudo C++ code is used for readability. Actual code is in machine language and directly written into memory.

- Let *descendants* of an element be elements that can be reached by 0-or-more-steps traversal of the data members whose type is a pointer to an element.
- Let *element variables* be data members of container or elements whose type is a pointer to an element.
- The actual parameter for p to call r from s should be a descendant of an element pointed to by a **private** data member of the container.
- The actual parameter for p to call r from r should be a descendant of either an element pointed by caller's p or a **private** data member of the container.
- If the container has a member function that modifies element variables, the new values should be either null pointers, pointers to descendants of objects pointed to by the container's element variables, or pointers to elements instantiated in the member function. If the member function is r, the new values may be pointers to descendants of an object pointed to by p.
- Elements should not have a member function that modifies element variables.
- The container should not export non-const pointers[6] to elements. That is, it should not have a member function to return them or to write them into globally accessible variables or memory addresses specified by arguments.

---

[6] To cast a pointer to const into a pointer to non-const is considered to be illegal, because it is not safe to modify an object through such a pointer[5].

```
// Before optimization
void StringList::add(char const *p) {
  StringLink *l=new StringLink(p);
  l->next=top; top=l;
}


// after optimization
void StringList::add(char const *p) {
  StringLink *l=new StringLink(p);
  <generate code and set its address to qqcode>
  l->next=top; top=l;
  l->next_code=top_code; top_code=qqcode;
}
```

Fig. 5. Member function add

- Elements should not have a member function to write non-const pointers to elements into globally accessible variables.

Program transformation by dataflow analysis, of course, allows cases where the actual parameter is a temporary variable that has a value of an element variable. Thus, the above conditions are general enough, not particularly hard to satisfy.

4.2 Generating Code Generators

This section describes a way to generate run-time code generators specialized to s and r. First, a code generator for s with first level inlining of r is generated (possible with an existing method) (Fig. 6(a)). Invariant data members of the container and the object pointed to by p are treated as run-time constants in the code generator. Here, optimizations across the s-r boundary are restricted. For example, constant propagation, code motion, and common subexpression elimination across the boundary are disabled[7]. This restriction is for reconfiguration of the data structure (code for r is replaced).

The resulting code generator is separated into that for s and r. The hole in s (after r's part is extracted) is filled with a code generator to invoke the code for the element pointed to by the actual parameter (Fig. 6(b)). An *address variable* described in the next section gives the code address.

Since r normally checks whether or not p is a null pointer, the code generator for r is duplicated: one optimized assuming p is a null pointer (generating code such as STRING1 in Fig. 4) and one optimized assuming p is not a null pointer (generating code such as NOSTRING in Fig. 4). The tail recursion part of r is replaced with a code generator to jump to the code for the element pointed to by the actual parameter (Fig. 6(c),(d)).

---

[7] Constant propagation from s to r is allowed.

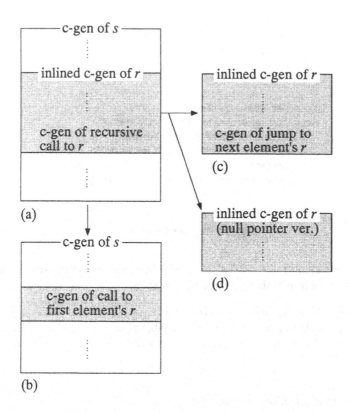

Fig. 6. Construction of Code Generators. ("c-gen" is short for "code generator".)

4.3   Using Code Generators and Generated Code

For each element variable, a new data member called an *address variable* is added to represent the address of the code for the element pointed to by the element variable. Member functions of the container are modified to assign the addresses of the code to address variables when the corresponding element variable is assigned.

The member function s is replaced with the following procedure:

– Check if the code for s is already generated. If not, invoke the code generator for s.
– Invoke the generated code for s.

Invocation of the non-null version of the code generator for r is embedded where the element is instantiated. Invariant data members of the instantiated object are treated as run-time constants. After the invocation, the address of the generated code is assigned to the address variable corresponding to the element variable to which the pointer to the instantiated object is assigned.

Since the null-pointer version of the code generator for r cannot use data members of an object pointed to by p, the resulting code must be independent of the elements. Thus invocation of the code generator is embedded into the constructor of the container. An assignment of the address of the generated code to an address variable is inserted where a null pointer is assigned to the corresponding element variable.

4.4   Deleting Generated Code

There are two types of code deletion: Deletion of code pointed to by an address variable is performed where the element pointed to by the corresponding element variable is deleted. The code of the null-pointer version of r generated in the constructor of the container is deleted in its destructor.

5   Evaluation

Four programs are compared to evaluate the method proposed in this paper. All are implementations of a linked list of strings described in Sect. 3. P1 is compiled normally. P2 is processed by a preprocessor [3] using an element-wise run-time optimization. P3 is the result of manual elimination of tail recursion and manual inlining of member function match. P4 is the result of the method proposed in this paper.

The target machine is an NEC PC-9821La10/8 (Pentium 100MHz, 23MBytes of memory, no secondary cache). The operating system is Windows 95. The compiler is BCC32A.EXE (a command line compiler of Borland C++ 4.5J). Compiler options are "-5 -O2 -vi" (optimized for Pentium, optimized for speed, inlining enabled).

Table 1 lists the execution times to search a list of ten 10-character-strings for both existing and non existing strings a million times. Each value is an average of ten executions.

Since the conventional method can optimize only member function match to its string and cannot eliminate function call overhead, speedup of P2 is very small. Manual inlining with tail recursion elimination (P3) works well, but since the inlined member function match is not optimized to its string, there is room for improvement. The proposed method allows all of these optimizations, and P4 shows the best performance.

program	exec. time (sec) / ratio	code gen. cost (msec)	code size (bytes)
P1. original	8.92 / 1.00	-	-
P2. conventional	8.76 / 0.98	2.69	1348
P3. manual inlining	3.33 / 0.37	-	-
P4. proposed	2.35 / 0.26	2.72	1251

Table 1. Linked list search results

program	string length * iteration				
	10*1000000	30*333333	100*100000	300*33333	1000*10000
P1. original	8.92	8.78	9.36	25.87	33.78
P4. proposed	2.35	2.84	11.04	22.56	27.71

Table 2. Effect of list length (execution time in seconds)

Table 1 also lists the run-time code generation costs and generated code sizes for P2 and P4. Several hundred times of iteration recoups the cost in P4. The generated code is several times as large as the original data structure. The code size may affect the performance when the list is very long.

Table 2 shows the execution times of P1 and P4 with various length of lists. P4 (proposed) is worse than P1 (original) in the case where the list length is 100. The generated code seems not to fit in the processor's instruction cache, while the original data structure fits in the data cache in this case. P4 performs better in all other cases.

6   Related Work

A number of research papers report methods of making it easy and practical to use run-time code generation. Among the methods are description languages and systems for run-time code generators [6] [7] [8], a run-time partial evaluation system for binary code [9], and automatic generation system of run-time code generators [10] [11] [12] [13] [14] [15]. But they deal with single functions, parts of functions, or program portions that can be arranged into single functions (e.g. using inlining).

One research paper mentions automatic generation of EDSs. Sect. 4.4 of [15] suggests that applying run-time partial evaluation can perform the same optimization as using the EDSs method. An example used there is a scheduler expressed in pseudo-code:

```
task *crt_tsk;

sched() {
  save_regs(crt_tsk);
  crt_tsk=crt_tsk->next;
  load_regs(crt_tsk);
  jmp_to(crt_tsk->PC);
}
```

Through specialization with respect to a given circular task queue, the resulting program is

```
sched_1() {
  save(FP1); save(FP4); // save registers for current task
  load(FP2); load(FP3); // load registers for following task
  jmp(0x....);
}
```

The second command in **sched**, which dereferences the pointer to the next task to be activated, is eliminated. Other commands are specialized with respect to the current and following tasks.

This result, however, though the example is overly simplified, is very far from the concept of the EDSs method. It does not make the data structure *self-traversing*. It fixes the order of the tasks. If a task's state changes from active to inactive, the context switch routines for both the previous and the following task should be changed. This is because the specialization views the whole queue as a run-time constant, like the conventional (trivial) type of optimizations described in Sect. 3. It disables the flexibility of linked structures. In the EDSs method, flexibility is preserved, and changing a task's state requires only the jump addresses in the switch code to be updated.

There may be some techniques, such as indirect jumps, that make the method in [15] flexible enough. But the programmer must carefully rewrite routines for updating the task switch routine. So it is not easy to use.

Since the method in this paper generates EDSs, the operation to update the links is light-weight. And this updating operation is automatically derived from the original source program.

7 Conclusion

This paper proposed a method of generating EDSs automatically from the source code in object-oriented languages, viewing them as an optimized implementation of a group of objects. With the help of an encapsulation mechanism of object-oriented languages, analyzing only the source code to implement the object group proves the method applicability. Generating code generators for EDSs is an extension of the conventional technique of generating run-time code generators. The preliminary evaluation shows that the proposed method is better than previous methods or the manual source level optimization.

The author will extend the method to deal with quasi-invariant data members. He will also implement a reverse operation of tail recursion elimination to accept iteration by loops.

References

1. Calton Pu, Henry Massalin, and John Ioannidis. The Synthesis kernel. *Computing Systems*, Vol. 1, No. 1, pp. 11–32, Winter 1988.
2. Henry Massalin. *Synthesis: An Efficient Implementation of Fundamental Operating System Services*. PhD thesis, Graduate School of Arts and Sciences, Columbia University, April 1992.
3. Nobuhisa Fujinami. Automatic Run-Time Code Generation in C++. In Yutaka Ishikawa, Rodney R. Oldehoeft, John V.W. Reynders, and Marydell Tholburn, editors, *LNCS 1343: Scientific Computing in Object-Oriented Parallel Environments. First International Conference, ISCOPE 97 Proceedings*, pp. 9–16, December 1997. Also appeared as Technical Report SCSL-TR-97-006 of Sony Computer Science Laboratory Inc.

4. Nobuhisa Fujinami. Determination of Dynamic Method Dispatches Using Runtime Code Generation. In Xavier Leroy and Atsushi Ohori, editors, *LNCS 1472: Types in Compilation. Second International Workshop, TIC'98 Proceedings*, pp. 253–271, March 1998. Also appeared as Technical Report SCSL-TR-98-007 of Sony Computer Science Laboratory Inc.
5. Bjarne Stroustrup, editor. *The C++ Programming Language, 2nd edition*. Addison Wesley, 1991.
6. Dawson R. Engler, Wilson C. Hsieh, and M. Frans Kaashoek. 'C: A language For High-Level, Efficient, and Machine-independent Dynamic Code Generation. In *Conference Record of POPL '96: The 23rd ACM SIGPLAN-SIGACT Symposium on Principles of Programming Languages*, pp. 258–270, January 1996.
7. Massimiliano Poletto, Dawson R. Engler, and M. Frans Kaashoek. tcc: A System for Fast, Flexible, and High-level Dynamic Code Generation. In *Proceedings of the SIGPLAN '97 Conference on Programming Language Design and Implementation*, pp. 109–121, June 1997.
8. Joel Auslander, Matthai Philipose, Craig Chambers, Susan J. Eggers, and Brian N. Bershad. Fast, Effective Dynamic Compilation. In *Proceedings of the SIGPLAN '96 Conference on Programming Language Design and Implementation*, pp. 149–159, May 1996.
9. Curtis Yarvin and Adam Sah. QuaC: Binary Optimization for Fast Runtime Code Generation in C. Technical Report UCB//CSD-94-792, University of California Berkeley, Department of Computer Science, 1994.
10. Peter Lee and Mark Leone. Optimizing ML with Run-Time Code Generation. In *Proceedings of the SIGPLAN '96 Conference on Programming Language Design and Implementation*, pp. 137–148, May 1996.
11. Mark Leone and Peter Lee. A Declarative Approach to Run-Time Code Generation. In *Workshop Record of WCSSS'96: The Inaugural Workshop on Compiler Support for System Software*, pp. 8–17, February 1996.
12. Brian Grant, Markus Mock, Matthai Philipose, Graig Chambers, and Susan J. Eggers. Annotation-Directed Run-Time Specialization in C. In *Proceedings of Workshop on Partial Evaluation and Semantics-Based Program Manipulation (PEPM'97)*, June 1997.
13. Brian Grant, Markus Mock, Matthai Philipose, Graig Chambers, and Susan J. Eggers. DyC: An Expressive Annotation-Directed Dynamic Compiler for C. Technical Report 97-03-03, Department of Computer Science and Engineering, University of Washington, 1997.
14. Charles Consel and François Nöel. A General Approach for Run-Time Specialization and its Application to C. Technical Report No. 946, INRIA/IRISA, July 1995.
15. Charles Consel, Luke Hornof, François Nöel, and Nicolae Volanshi. A Uniform Approach for Compile-time and Run-time Specialization. Technical Report No. 2775, INRIA, January 1996.

Improving Cache Utilization of Linear Relaxation Methods: Theory and Practice

Federico Bassetti[1], Kei Davis[1], Madhav Marathe[1],
Dan Quinlan[2], and Bobby Philip[2]

[1] Los Alamos National Laboratory, Los Alamos, NM 87545, USA
[2] Lawrence Livermore National Laboratory, Livermore, CA, USA

Abstract. Application codes reliably achieve performance far less than the advertised capabilities of existing architectures, and this problem is worsening with increasingly-parallel machines. For large-scale numerical applications, stencil operations often impose the greater part of the computational cost, and the primary sources of inefficiency are the costs of message passing and poor cache utilization. This paper proposes and demonstrates optimizations for stencil and stencil-like computations for both serial and parallel environments that ameliorate these sources of inefficiency. Additionally, we argue that when stencil-like computations are encoded at a high level using object-oriented parallel array class libraries, these optimizations, which are beyond the capability of compilers, may be automated. The automation of these optimizations is particularly important since the transformations represented by cache based optimizations can be unreasonably complicated by the peculiarities which are architecture specific. This paper briefly presents the approach toward the automation of these transformations.

1 Introduction

Modern supercomputers generally have deep memory hierarchies comprising a small fast memory (registers) and increasingly large, increasingly slow memories. Five such levels exist for the Los Alamos National Laboratory ASCI Blue machine: each processor sees a register file, level one (L1) and level two (L2) caches, main memory, and remote main memory. Future architectures may rely on yet deeper memory hierarchies.

It is clear that the realization of modern supercomputer potential relies critically on program-level cache management via the staging of data flow through the memory hierarchy to achieve maximal data re-use when resident in the upper memory levels. In general it is important to utilize the spatial- and temporal locality of reference in a problem. Data exhibits temporal locality when multiple references are close in time; spatial locality when nearby memory locations are referenced. This is important because memory access time (from the same level) may not be uniform, e.g. in loading a cache line or memory page.

Much research has been devoted to studying pragmatic issues in memory hierarchies [1,2,3,4,5,6] but relatively little to the basic understanding of algorithms

S. Matsuoka et al. (Eds.): ISCOPE'99, LNCS 1732, pp. 25–36, 1999.

and data structures in the context of memory access times. Recently hierarchical memory models have been proposed [7,8,9,10,11] and efficient algorithms for a number of basic problems have been proposed under these models.

2   Background and Relationship to Previous Work

Relaxation is a well-known technique in the solution of many numerical problems. We recall the ideas from [16,14] on graph covers based methods for efficient implementation of linear relaxation problems. A prototypical problem in this context can be described by a graph $G(V, E)$ in which each vertex $v \in V$ contains a numerical value x_v which is iteratively updated. At iteration t, of the linear relaxation process, the state of v is updated using the equation $x_v^{(t)} = \sum_{(u,v) \in E} A_{uv}^{(t)} x_u^{(t-1)}$ where $A_{uv}^{(t)}$ denotes the relaxation weight of the edge (u, v). This equation may written using matrix notation as $x^{(t)} = A^{(t)} x^{(t-1)}$, where $x^{(t)} = < x_1^{(t)}, x_2^{(t)}, \ldots, x_{|V|}^{(t)} >$. The goal of the relaxation is to compute a final state vector $x^{(T)}$ given an initial state vector $x^{(0)}$. In this paper, we will assume that A does not change over time.

A simple and effective way to compute relaxation is to update the state of each vertex in the graph according to the equation given above and then repeat this step until we obtain $x^{(T)}$. This is inefficient in today's computing architectures where the memory system typically has two or more levels of cache between the processor and the main memory. The time required to access a data value from memory depends crucially on the location of the data; the typical ratios of memory speeds are $1 : 10 : 100$ between the first level cache, second level cache and the main memory. As CPUs get faster these processor-memory gaps are expected to widen further. As a result it is important to design transformations for applications (in this case relaxation methods) that take into account the memory latency. Such algorithms seek to stage the computation so as to exploit spatial and temporal locality of data. Common compiler transformations such as blocking mainly address spatial locality. As will be shown, significant performance can be obtained by addressing temporal locality.

These optimizations form a specific instance of a more general optimization that originates in older out-of-core algorithms, main memory in this case is treated as slower storage (out-of-core). In this case we treat the instance of a stencil operation on a structured grid as a graph and define covers for that graph. The covers define localized regions (blocks) and form the basis of what we will define to be τ-*neighborhood-covers*. The general idea behind applying this transformation to structured grid computations is to cover (or *block* or *tile*) the entire grid by smaller sub-grids, solving the problem for each subgrid sequentially. Let n be the number of grid points in a two dimensional square domain, and A be a $\sqrt{n} \times \sqrt{n}$ grid. Let the size of L_1 cache be M. Now consider solving the linear relaxation problem for a subgrid S of size $k \times k$. In the first iteration, all the points in S can be relaxed. After the second iteration, points in the grid of size $(k-2) \times (k-2)$ have the correct value. After continuing the procedure for τ

iterations, it follows that a subgrid of size $(k-2\tau) \times (k-2\tau)$ has been computed up to τ time steps. Thus we can now cover the $\sqrt{n} \times \sqrt{n}$ by $\frac{\sqrt{n} \times \sqrt{n}}{(k-2\tau) \times (k-2\tau)}$ which can be rewritten as $\frac{n}{(k-2\tau)^2}$. The total number of loads into the L_1 cache is no more than $k^2 \frac{n}{(k-2\tau)^2}$. In order to carry out the relaxation for T time units, the total number of L_1 loads is no more than $k^2 \frac{n}{(k-2\tau)^2} \frac{T}{\tau}$. By setting $k = \sqrt{M}$, and $\tau = \sqrt{M}/4$, we can obtain an improvement of $O(\sqrt{M})$ in the number of L_1 loads over a naive method (which would take $O(Tn)$ time).

In practice, for the solution of elliptic equations using multigrid methods on structured grids, τ is typically a small constant between 2 and 10 and thus the asymptotic improvement calculated above does not directly apply; nevertheless the analysis shows that we can get constant factor improvements over the naive method in terms of the memory access. The naive method we refer to here is the blocking introduced by the compiler. Of course less efficient methods could be proposed (e.g. local relaxation methods) which would permit significantly larger values of τ, but these less efficient solution methods are not of practical interest. The idea can be extended in a straightforward way to the cases when the value at a grid point is calculated by using not only the neighboring values but all neighbors that are a certain bounded distance away.

In the above calculations we have chosen to recalculate the values on the boundary of each subgrid since this does not give us any additional asymptotic disadvantage. In practice, however, these values are maintained in what we call *transitory* arrays. This introduces additional complications, namely the need to ensure that the transitory arrays are resident in the L_1 cache. This implies that for a 2-dimensional case we need to maintain arrays with total size $4k + 4(k-2) + \cdots + 4(k-2(\tau-1)) \sim O(k^2)$. Although asymptotically this is still the same as the size of our subgrid, the constants need to be carefully calculated in order to determine the best parameter values. The memory requirements for the transitory array could be reduced slightly by lexicographically ordering the subgrids in terms of the grid coordinates. This allows us to throw away the left and the top boundary of a subgrid.

These ideas can be extended to general graphs that capture the dependency structure. For this purposes, given a graph $G(V, E)$ let $N^\tau(v) = \{w \mid d(w, v) \leq \tau\}$. Here $d(w, v)$ denotes the distance in terms of number of edges from v to w.

Definition. τ-**Neighborhood Cover**: Given a graph $G(V, E)$, a τ-neighborhood cover $\mathcal{G} = \{G_1, \ldots, G_k\}$ is set of subgraphs G_i, $1 \leq i \leq k$, such that $\forall v \in V$, there exists a G_i with the property that $N^\tau(v) \subseteq V_i$.

Given a graph $G(V, E)$, a *sparse* τ-neighborhood cover for G is a τ-neighborhood cover with the additional properties $\forall i$, $1 \leq i \leq k$, $|G_i| \leq M$, and $k = O(|E|/M)$. Note that a vertex v can belong to more than one subgraph G_i but for the purposes of computing its value there is at least one subgraph G_v that may be regarded its "home", meaning that the state $x_v^{(\tau)}$ at a vertex v at time τ can be calculated by just looking at the graph G_v. Existence of sparse neighborhood covers for a graph G imply the existence of memory efficient algorithms for linear relaxation on G. Specifically, given a graph G with a sparse τ-neighborhood cover, it is possible to complete T steps of relaxations using $O(\frac{T}{\tau}|E|)$ loads in the L_1 cache as opposed to $O(T|E|)$ loads in the naive implementation. This can be seen by observing that τ steps of the linear relaxation can be completed using no more than $O(M)O(|E|/M)$ L_1 loads. Certain well known classes of graphs (e.g. graphs for finite difference computational grids) have sparse neighborhood covers.

The concept of graph covers closely resembles the concept of *tiling* that has been extensively used in the past to reduce the synchronization and memory access costs. Tiling transformation is one of the primary transformations used to improve data locality. Tiling generalizes two well-known transformations, namely, *strip-mine and interchange* and *unroll and jam*. Graph covers can be thought of as a generalization of tiling. In order to see this, we recall some basic definitions from compiler theory [12,15] to which we refer the the reader, and to the referenced cited therein, for more details on this topic.

Iteration Space. A set of n nested for loops are represented as a polytope $\mathbf{Z}^b$ with the bounds of the polytope corresponding to the bounds placed on the loops. Each iteration now corresponds to a point in $\mathbf{Z}^n$. We think of these points as vertices of a graph called the *dependence graph*. The vertex is identified by a vector $p = (p_1, \ldots, p_n)$, where p_i is the value of the ith loop index in the nest while counting from outermost to the innermost loop. It is easy to see that each axis represents a loop and each vertex represents an iteration that is performed. With this notation, it is clear that an iteration $p = (p_1, \ldots, p_n)$ has to be executed before another iteration $q = (q_1, \ldots, q_n)$ iff p is *lexicographically smaller* than q^1 (denoted $p \prec q$). This intuition can be captured via the notion of *distance and dependence vectors or edges*. Viewing the points corresponding to the iterations in $\mathbf{Z}^n$ as vertices of a graph, we have a directed edge from a vertex p to a vertex q with label d if $p \prec q$ and $d = q - p$. The directed graph (called the dependence graph) assigns a partial order to the vertices corresponding to the iterations; the important point is that any complete ordering of the original vertices that respect the partial ordering is a feasible schedule for executing of the iterations.

Tiling. In general tiling maps a n-deep nested loop structure into a $2n$ deep nested loop structure with only a small fixed number of iterations. Graphically, tiling can be viewed as way to break the iteration space into smaller blocks with the following conditions:

1. The blocks are disjoint.
2. The individual points in each block and the block themselves are ordered in such a way so as to preserve the dependency structure in the original iteration space; i.e. if $p \prec q$, then p is executed before q. Stated another way, the complete ordering for scheduling the execution of the iterations should respect the partial ordering given by the dependence graph.
3. The blocks are by and large rectangular, excepting when the iteration spaces are themselves of a different shape (e.g. trapezoidal, triangular).

---

[1] $p \prec q$ iff $1 \leq i \leq n$, $p_i \leq q_i$.

4. Typically, if the original problem had a set of n loop indices, the transformation yields $2n$ indices (one new index for each old index) such that the bounds on original n indices do not depend on each other, but rather depend only on the corresponding new index created in the process of the transformation.

In contrast, graph covers create a set of smaller subgraphs and yield a transformed instance in which *either* certain iterations are revisited (i.e. steps at those times are recalculated) thus creating overlapping tiles, *or* it stores the intermediate values and avoids the extra computation. Intuitively, tiling based on graph covers creates trapezoidal tiles, and uses temporary storage to make sure that the dependency graph partial order is maintained.

In this paper Jacobi relaxation is used as an example relaxation code, using a two-dimensional array and a five-point stencil (diagonal elements are excluded). Such computations appear as parts of more sophisticated algorithms as well (e.g. multigrid methods). A single iteration or *sweep* of the stencil computation is of the form

```
for (int i=1; i != I-1; i++)
 for (int j=1; j != J-1; j++)
  A[i][j] = w1*A[i-1][j] + w2*A[i+1][j] + w3*A[i][j-1] + w4*A[i][j+1];
```

Typically several sweeps are made, with A and B swapping roles to avoid copying. It is easy to see that the dependence graph of this loop is not strict. Thus the tiling theory developed elsewhere [12,13,15] cannot be applied directly to yield the type of transformations we obtain.

In typical serial or parallel C++ array class library syntax the statement is

```
A(I,J) = w1*A(I-1,J) + w2*A(I+1,J)+ w3*A(I,J-1) + w4*A(I,J+1);
```

While the transformations we will detail are general, it is the prototypical array class syntax that is targeted for such optimizations.

2.1   Summary of Results

The main goal of this paper is to perform empirical analysis of the graph covers based transformation to improve the performance of linear relaxation problems. The basic theory of graph covers is not new; these ideas have been discussed in a number of earlier theoretical papers (see [14,5,16] and the references therein). In contrast, we are not aware of extensive experimental evaluation of graph covers based techniques in the context of improving linear relaxation algorithms. The results in [5,16] are discussed in the context of out-of-core algorithms; while our main focus is on designing algorithms that use the cache effectively. Moreover, as mentioned earlier, the asymptotic results in [5,16] can not be applied directly since the number of iterations in practice is typically a small constant. Thus our focus was to empirically find out what constant factor improvements can be obtained using the graph covers approach. For this, the implementations required special care. For instance, while the asymptotic analysis ignores the effect of recalculating the boundary values due to small constant overhead, our implementations needed to take this into account. As mentioned earlier, this implied maintaining transitory arrays for storing intermediate values of the boundary elements. Thus the implementation had to take into account the computation time versus space tradeoffs to obtain efficient implementations. The results obtained in this paper are only for Jacobi relaxation; extension to other relaxation methods will be the focus of subsequent papers. The results obtained in this paper are a step towards empirically showing that graph cover based methods can be used for design efficient algorithms by exploiting temporal locality inherent in certain linear relaxation problems. Section 3 provides further details justifying the claim. We have also started integrating our results within an objected oriented framework called ROSE for automating these transformations (see Sect. 4).

3   Performance Results

Our technique is compared to a naive implementation relying on an optimizing compiler. The performance data were gathered on SGI Origin 2000 system comprising 128 MIPS R10000 processors running at 250 Mhz. Each processor has a split L1 cache with 32 Kbytes for instructions and 32 Kbytes for data. L2 is a unified 4 Mbytes. L1 and L2 are both 2-way associative. The ratio of L1 and L2 line size 1:4. The system has 250 Mbytes per processor. The test codes were compiled using the MIPSpro C++ with O3 optimization. Performance data were gathered via the hardware performance monitors available on the processors. The test code is a Jacobi iterative solver based on the code fragment above.

3.1   Single Processor Performance

Ideal performance is obtained when the entire working set fits in primary cache; this is shown in Figures 1, 2, and 3. Three metrics are used: *cycles* to describe the overall execution; *primary cache data misses* to describe the impact of memory penalty, and *flops* to describe how efficiently, compared to the ideal, the test code performs. All results are based on the average cost of one iteration. Figure 1 shows that the cost of each iteration generally decreases as the number of iterations increases: all misses are first-time-load misses—once a datum is loaded into cache it remains resident for the duration of the computation—as shown in Figure 2, so increasing the number of iterations amortizes the cost of first-time-load misses.

The vast majority of scientific applications carrying out meaningful computations have a working set that is many times larger than any cache. Figure 4 shows the performance of the naive implementation of the test. The cost of each iteration is on average the same regardless of the number of iterations. Again, this is a consequence of the cache miss behavior; Figure 5 confirms that every iteration entails the same number of misses regardless of the number of iterations. The efficiency is less than half of the idealized cache-resident version as shown in Figure 6.

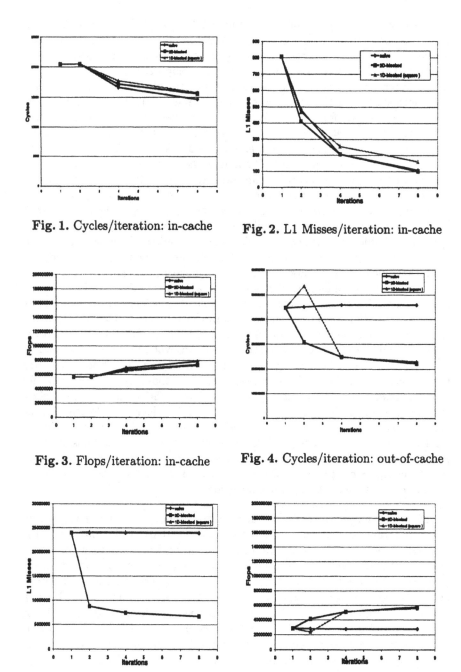

Fig. 1. Cycles/iteration: in-cache

Fig. 2. L1 Misses/iteration: in-cache

Fig. 3. Flops/iteration: in-cache

Fig. 4. Cycles/iteration: out-of-cache

Fig. 5. L1 Misses/iteration: out-of-cache

Fig. 6. Flops/iteration: out-of-cache

Most scientific applications have inherent locality; this is a simple consequence of array-based computation. Modern compilers are able to detect and exploit some locality. Compilers exploit locality only within one iteration; reuse is minimal or nil between iterations. Exploiting temporal locality with the technique presented in the previous sections demonstrably reduces the difference in performance from the ideal cache-resident version. Figures 4, 5, and 6 give performance results for optimized version of the test code. In all the figures *2D-blocked* and *1D-blocked (square)* represent cache blocking performed using, respectively, a two dimensional tile and a one dimensional tile (a stripe) for a square two dimensional problem. The figures show that the *2D-blocked* version performs better than the others. Quantitatively, it is possible to observe a factor of two improvement compared to the naive implementation. The explanation for such differences in performance is in the number of misses, as show in Figure 5. While in the naive version the number of misses is the same per iteration, the optimized 2D-blocked one amortizes the number of misses. This says that in the optimized code the number of misses performed during the first iteration are the significant ones, besides those the number of misses performed is significantly less, such that the total count of misses could appear, relatively, independent of the number of iterations executed. There are noticeable similarities between the performance of the 2D-blocked version and the ideal (cache resident) version: the cost in cycles per iteration decreases when increasing the number of total iterations; the number of misses per iteration decreases when increasing the number of total iterations; the efficiency, in flops per second, increases when increasing the number of iterations. Nevertheless, it appears that there is still room for further investigation and possible improvement since the relative flop rates show that there is still a difference between ideal and 2D-blocked performance. The performance obtained with the 1D-blocked version, overall, can be considered similar to the 2D-blocked one with the exception of primary cache misses. For a large problem size, a tile is a stripe that cannot fit in L1 or L2 cache. However, the stripe approach exploits reuse for L2 cache in proportions good enough to macth performance of the *2D-blocked* version.

The performance figures show that the transformation is in fact an optimization; validation on different platforms is needed. Of particular interest is the comparison between 2D-blocked and 1D-blocked implementations. In theory, a 2D-blocked implementation should outperform an equivalent 1D-blocked one. In practice, it appears that this theoretical difference in performance is not noticeable. With the implementation of the temporal locality optimization the loop structure, if compared to the naive one, has been significantly modified: loops have been added and a temporary data structure is needed. We believe that while the technique presented has a great impact on the memory access pattern, improving by factor of 5-10 the number of misses, overall performance improves by a smaller factor because of the a more complicated loop structure that makes pipelining less efficient. This effect might be the limiting factor for the temporal blocking technique. The maximum achievable improvement factor is still under investigation.

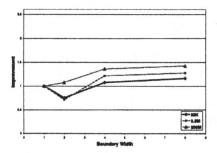

Fig. 7. Improvement by widening boundaries on 8 processors. Problem size is given in number of points.

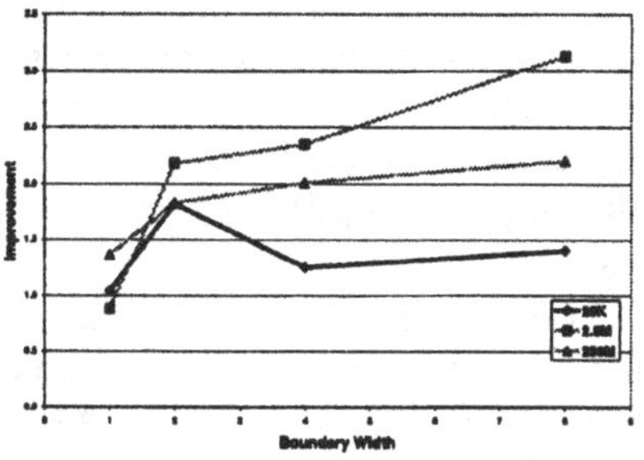

Fig. 8. Improvement by the optimized implementation on 8 processors. Problem size is given in number of points.

3.2 Multiple Processors

In a parallel environment the arrays are typically distributed across multiple processors. To avoid communication overhead for calculations near the boundaries of the segments of the the arrays, space is traded for time by creating *ghost boundaries*—read-only sections of the other processors' array boundaries. In the parallel case the C++ array statement transparently abstracts the distribution of data, the parallelism in the execution of the operations on the data, and the message passing required along the partitioned edges of the distributed data.

The basic idea is to trade computation, which is relatively cheap, for communication, which is expensive, to reach an optimal compromise, by increasing the amount of data at block boundaries, making possible multiple iterations over the data before communication is required. Figure 7 demonstrates that this strategy is beneficial to overall performance. This technique is described elsewhere [17].

In this work we use the message passing paradigm, using the *native* implementation of MPI. Figure 8 shows improvement achieved by the optimized code. In almost all of the cases studied the optimized version is between two and three times faster than the naive implementation. The improvement that has been demonstrated for single processor is maintained, undegraded, in the multiprocessor case. We have verified this effect up to 64 processors.[2]

4   Automating the Transformations

To be of practical use an optimizing transformation must be automated. In the context of C++ array classes it does not appear possible to provide this sort of optimization within the library itself because the applicability of the optimization is context dependent—the library can't know how its objects are being used.

---

[2] Since a multidimensional space graph is needed to have a complete picture only a portion of those are presented.

Two mechanisms for automating such optimizations are being actively developed: the use of *expression templates* by others, and a source-to-source transformation system (a pre-processor), which we are currently developing. The ROSE preprocessor is a mechanism for C++ source-to-source transformation, specifically targeted at optimizing (compound) statements that manipulate array class objects.

ROSE is a programmable source-to-source transformation tool for the optimization of C++ object-oriented frameworks [18]. In our initial work we target the OVERTURE Framework (www.llnl.gov/casc/Overture), a parallel object-oriented C++ framework for solving partial differential equations associated with computational fluid dynamics applications within complex moving geometries. While we have specific goals for this work within OVERTURE, ROSE is applicable to any other object-oriented framework.

A common problem within object-oriented C++ scientific computing is that the high level semantics of abstractions introduced (e.g. parallel array objects) are ignored by the C++ compiler. Classes and overloaded operators are seen as unoptimizable structures and function calls. Such abstractions can provide for particularly simple development of large scale parallel scientific software, but the lack of optimization greatly affects performance and utility. Because C++ lacks a mechanism to interact with the compiler, elaborate mechanisms are often implemented within such parallel frameworks to introduce complex template-based or runtime optimizations such as runtime dependence analysis, deferred evaluation, and runtime code generation.

ROSE is a system for implementing preprocessors that read the user's application source code and output highly optimized C++ code. ROSE helps the framework developer define framework-specific (or application specific) grammars (more specifically a hierarchy of grammars). Multiple program trees are built, one for each grammar. The traversal of the much simpler program trees represented by the higher level grammars (as opposed to that of the more complex C++ program tree) permits the identification of locations where transformations are to be applied. The final modified program tree is then unparsed to generate C++ source code. The source code transformations can readily exploit knowledge of the architecture, parallel communication characteristics, and cache architecture in the specification of the transformations. These transformations may range from serial loop optimizations to parallel message passing optimizations, threading could alternatively be automated with such transformations (where identified using the framework's parallel semantics). To simplify debugging of the optimized application code, the output of the unparsed program tree matches the format of the input application program.

Since the framework's semantics are preserved and the transformations are introduced optionally, where possible, the entire use of the preprocessor is optional within the development process of the application. Experience shows preprocessing to require time of the same order as compilation. The approach is thus fast, unobtrusive, and practical, and more programmable than template based approaches. More sophisticated transformations are possible because the context of the use of the abstractions (objects) relative to one another can be seen, and the semantics of these objects is known. Dependence analysis, for example, is relatively simple when leveraged with the semantics of array abstractions.

ROSE is designed using several other tools. We use the EDG C++ front-end and the SAGE source code restructuring tool [19,20]. ROSE exists as a layer on top of SAGE (an open interface to the C++ program tree provided though the EDG front-end), while SAGE exists as a layer on top of EDG. EDG is a commercial C++ front-end, providing us with an implementation of the full C++ language (as complete as is available today). By design, we leverage many low level optimizations provided within modern compilers while focusing on higher level optimizations. Key design features of ROSE include (i) decoupling optimization from compilation, (ii) use of hierarchies of grammars, (iii) provision of a simple mechanism for specifying transformations, (iv) not requiring semantic inference/program analysis. Finally, because ROSE is based ultimately (through SAGE) upon the EDG C++ front-end, the full language is made available, consistent with the best of the commercial C++ compilers.

5 Conclusions and Future Work

We posit that current optimizations for stencil-based applications are inadequate for which desirable optimizations exist that cannot reasonably be expected to be implemented by a compiler. One such optimization for cache architectures has been detailed and demonstrated to give a factor of two improvement in performance in a realistic setting. The transformation is language-independent, though it is demonstrated only for C code.

The use of object-oriented frameworks is a powerful tool, but performance is generally less than that of FORTRAN 77. We expect that in the future one will use such object-oriented frameworks because they represent *both* a higher-level, simpler, and more productive way to develop large-scale applications *and* a higher performance development strategy. We expect higher performance because the representation of the application using the higher level abstractions permits the use of new tools such as the ROSE II optimizing preprocessor.

We are currently investigating a number of extensions of our work reported here. This includes, generalization to higher dimensions, effect of different compilers and register allocation schemes on the performance, extension of ideas to multiple levels of hierarchy by a recursive application of the idea and use of threads and OpenMP for parallel implementation. Finally, we note that recently other authors have also investigated similar methods for improving the cache utilization of certain iterative algorithm. [21,22].

References

1. A. Aggarwal and J. Vitter. The I/O complexity of sorting and related problems. In *Comm. of the ACM (CACM)*, 1988. pp. 1116-1127.
2. P.J. Denning. Virtual memory. In *ACM Computing Surveys*, 1970. pp. 153-189.

3. P.C. Fischer and R.L. Probert. Storage reorganization techniques for matrix computation in a paging environment. In *CACM*, 1979. pp. 405-415.
4. R.W. Floyd. Permuting information in idealized two level storage. In R.E. Miller and J.W. Thatcher, eds., *Complexity of Computer Computations*. Plenum Press, 1972. pp. 105-109.
5. J. W. Hong and H.T. Kung. I/o complexity: The red blue pebble game. In *Proc. 13th ACM Symp. Th. Comp.*, 1981. pp. 326-333.
6. D.R. Shultz R.L. Mattson, J. Gacsei and I.L. Traiger. Evaluation techniques for storage hierarchies. In *IBM Systems J.*, pages pp. 78-117, 1970.
7. A.K. Chandra A. Aggarwal and M. Snir. Hierarchical memory with block transfer. In *Proc. 28th IEEE Symp. on Foundations of CS*, 1987. pp. 204-216.
8. A.K. Chandra A. Aggarwal, B. Alpern and M. Snir. A model for hierarchical memory. In *Proc. 19th ACM Symp. Th. Comp.*, pages pp. 305-314, 1987.
9. B. Alpern, L. Carter, E. Feig, and T. Selker, Uniform Memory Hierarchy Model of Computation. In *Proc. 31st IEEE Symp. on Foundations of CS*, 1990.
10. J. Vitter, Efficient Memory Access in Large Scale Computations. invited talk at *Symp. Theoretical Aspects of Computer Science, (STACS)* 1993.
11. J. Vitter, External Memory Algorithms and Data Structures. in *Proc. DIAMCS Series on Discrete Mathematics and Theoretical Computer Science*. 1998.
12. M. Wolf. *Improving Locality and Parallelism in Nested Loops*. PhD thesis, Department of Computer Science, Stanford University, 1992.
13. M. Wolfe. *High Performance Compilers for Parallel Computing*. Addison Wesley, 1996.
14. B. Awerbuch B. Berger, L. Cowen and D. Peleg. Near-linear cost sequential and distribured constructions of sparse neighborhood covers. in *Proc. 34th Symp. on Foundations of CS*. pp. 638-647, Palo Alto, CA, Nov. 1993.
15. S. Carr. *Memory Hierarchy Management*. PhD thesis, Rice University, 1992.
16. C. Leiserson, S. Rao and S. Toledo. Efficient out-of-core algorithms for linear relaxations using blocking covers. In *J. Comp. and System Sci.*. 54(2):332-344, April 1997.
17. F. Bassetti, K. Davis, M. Marathe, and D. Quinlan. Loop transformations for performance and message latency hiding in parallel object-oriented frameworks. In *Int. Conf. Par. Distr. Proc. Techniques and Applications*, 1998.
18. K. Davis and D. Quinlan, ROSE: An Optimizing Code Transformer for C++ Object-Oriented Array Class Libraries. *Workshop on Parallel Object-Oriented Scientific Computing (POOSC'98)*, LNCS 1543, Springer-Verlag, 1998.
19. B. Francois et. al. Sage++: An object-oriented toolkit and class library for building fortran and c++ restructuring tools. In *Proc. 2nd Object-Oriented Numerics Conf.*, 1994.
20. Info available at: http://www.edg.com.
21. J. Mellor-Crummey, D. Whalley and K. Kennedy. Improving Memory Hierarchy Performance for Irregular Applications. In *Proc. 13th ACM-SIGARCH Int. Conf. Supercomputing*. Greece, 1999.
22. C. Weis, W. Karl, M. Kowarschik and U. Rüde. Memory Characteristics of Iterative Methods. to appear in *Supercomputing'99*. Portland, Nov. 1999.

An Object-Oriented Framework for Parallel Simulation of Ultra-large Communication Networks*

Dhananjai Madhava Rao and Philip A. Wilsey

University of Cincinnati, Cincinnati, OH, USA

Abstract. Communication networks have steadily increased in size and complexity to meet the growing demands of applications. Simulations have been used to model and analyze modern communication networks. Modeling and simulation of networks involving thousands of nodes is hard due to their sheer size and complexity. Complete models of the ultra-large networks need to be simulated in order to conduct in-depth studies of scalability and performance. Parallel simulation techniques need to be efficiently utilized to obtain optimal time versus resource tradeoffs. Due to the complexity of the system, it becomes critical that the design of such frameworks follow well established design principles such as object oriented (OO) design, so as to meet the diverse requirements of portability, maintainability, extensibility, and ease of use. This paper presents the issues involved in the design and implementation of an OO framework to enable parallel simulation of ultra-large communication networks. The OO techniques utilized in the design of the framework and the application program interfaces needed for model development are presented along with some experimental results.

1 Introduction

Communication networks coupled with the underlying hardware and software technologies have constantly increased in size and complexity to meet the ever growing demands of modern software applications. The Internet, a global data network, now interconnects more than 16 million nodes [1]. Modeling and studying today's networks with their complex interactions is a challenging task [1,2]. Simulation has been employed to aid the study and analysis of communication networks [2]. Parallel simulation techniques have been employed in large simulations to meet the resource requirements and time constraints. The network models are critical components of simulation analyses that should reflect the actual network sizes in order to ensure that crucial scalability issues do not dominate during validation of simulation results [1]. Many networking techniques that work fine for small networks of tens or hundreds of computers may become impractical when the network sizes grow [1]. Events that are rare or that do not

---

* Support for this work was provided in part by the Defense Advanced Research Projects Agency under contract DABT63–96–C–0055.

S. Matsuoka et al. (Eds.): ISCOPE'99, LNCS 1732, pp. 37–48, 1999.

even occur in toy models may be common in actual networks under study [1]. Since today's networks involve a large number of computers ranging from a few thousands to a few million nodes, modeling and simulating such ultra-large networks is necessary.

Parallel simulation of ultra-large networks involves complex interactions between various components and places high demands on system resources — especially system memory. Hence, the data structures used in the system must be optimally designed with robust interfaces to enable efficient exchange of control and data in the distributed system [3]. To ease modeling, verification, simulation, and validation of ultra-large networks, well defined design principles such as object oriented (OO) design must be employed. In order to meet these diverse needs, an OO framework for simulation of ultra-large communication networks was developed. The framework is built around WARPED [3], an optimistic parallel simulation kernel. The framework is implemented in C++ and utilizes the OO techniques of inheritance, virtual functions, and overloading to develop efficient and robust interfaces.

The issues involved in the design and implementation of the framework along with a compendium of OO techniques used are presented in this paper. In Sect. 2 some of the related work in large scale network simulations are presented. Section 3 contains a brief description of WARPED. A detailed description of the framework and the application program interface (API) along with the OO techniques employed in their design and implementation are presented in Sect. 4. Some of the experiments conducted using the framework are presented in Sect. 5. Section 6 contains some concluding remarks.

2 Related Work

Simulation of large scale network models has received considerable attention in the past. Various combination of techniques have been used to improve the capacity and efficiency of large scale network simulations. Huag *et al* present a novel technique to selectively abstract details of the network models and to enhance performance of large simulations [4]. Their technique involves modification of the network models in order to achieve abstraction [4]. Premore and Nicol present issues involved in development of parallel models in order to improve performance [5]. In their work, they convert source codes developed for *ns*, a sequential simulator to equivalent descriptions in Telecommunications Description language (TeD) to enable parallel simulation [5]. Coupled with meta languages (such as TeD) [5], parallel network libraries and techniques to transparently parallelize sequential simulations have been employed [4]. Relaxation and even elimination of synchronization, a large overhead in parallel simulations, has been explored [6,7]. The relaxation techniques attempt to improve performance at the cost of loss in accuracy of the simulation results [6]. Fall exploits a combination of simulation and emulation in order to study models with large real world networks [8]. This method involves real time processing overheads and necessitates detailed model development.

In this paper, we present a technique for collating similar object descriptions to reduce and regulate the memory consumptions of the model and the simulation kernel. OO techniques have been employed to provide a robust API to ease model development and insulate the users from the underlying details. The API is similar to WARPED, the underlying simulation kernel. This enables other networking models, such as active networking models [9], to seamlessly interoperate with the framework. The paper also demonstrates effectiveness of the framework to enable ultra-large network simulations by providing experimental results.

3 Background

The framework for ultra-large networks simulation is built around the WARPED simulation kernel. WARPED [3] is an optimistic parallel discrete event simulator and uses the Time Warp mechanism for distributed synchronization. In WARPED, the logical processes (LPs) that represent the physical processes being modeled are placed into groups called "clusters". The clusters represent the operating system level parallel processes constituting the simulation. LPs on the same cluster directly communicate with each other without the intervention of the messaging system. This technique enables sharing of events between LPs, which considerably reduces memory overheads. Communication across cluster boundaries is achieved using MPI. LPs within a cluster operate as classical Time Warp processes; even though they are grouped together, they are not required to operate in time lockstep. A periodic garbage collection technique based on Global Virtual Time (GVT) is used. WARPED presents a simple and robust OO application program interface (API). Control is exchanged between the application and the simulation kernel through cooperative use of function calls. Further details on the working of WARPED and information on its API are available in the literature [3].

4 The Ultra-large Scale Simulation Framework (USSF)

The ultra-large scale simulation framework was developed to ease modeling and simulation of large communication networks. As shown in Fig. 1, the primary input to the framework is the topology to be simulated. The syntax and semantics of the input topology is defined by the Topology Specification Language (TSL), which provides simple and effective techniques to specify hierarchical topologies [9]. The topology is parsed into an OO Intermediate Format (TSL-IF). Static analysis is performed on the intermediate format to extract and collate common object definitions. The analyzed TSL-IF is then used to generate an optimal simulatable network topology. The current implementation of USSF, in conjunction with WARPED and the generated code, is in C++. The generated topology includes code to instantiate the necessary user defined modules that provide descriptions for the components in the topology. The generated code is

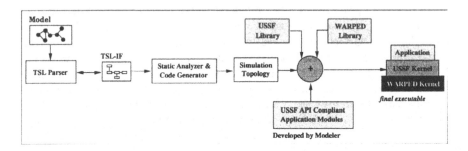

Fig. 1. Overview of USSF

compiled along with the USSF library, the WARPED library, and the application program modules to obtain the final simulation executable. The following sections describe the various components in detail.

4.1 Topology Specification Language (TSL)

The topology of the network to be simulated is provided to the framework in Topology Specification Language (TSL) [9] syntax. A TSL specification consists of a set of topology specifications. Each topology specification consists of three main sections, namely; (i) the *object definition section* that contains the details of the modules to be used to simulate the topology; (ii) the *object instantiation section* that specifies the various nodes constituting the topology; and (iii) the *netlist section* that defines the interconnectivity between the various instantiated components. An optional *label* may be associated with each topology. The label may be used as an object definition in subsequent topology specifications to nest a topology within another. In other words, the labels, when used to instantiate an object, result in the complete topology associated with the label to be embedded within the instantiating topology. Using this technique, a simple sub-net consisting of merely ten nodes can be recursively used to construct a network with six levels of hierarchy to specify a million (10^6) node network.

The input topology configuration is parsed by a TSL-Parser into an OO TSL Intermediate Format (TSL-IF). The current implementation of the TSL-Parser is built using the Purdue Compiler Construction Tool Set (PCCTS) [10]. TSL-IF is designed to provide efficient access to related data from the various TSL sections. Each component of the grammar is represented by a corresponding TSL-IF node (or object). Every node in TSL-IF is built from the same basic building block; in other words, every node is derived from the same base class. The intermediate format is composed by filling in the appropriate references to the different nodes generated by the parser. Since composition is achieved via base class references, each node can refer to another node or even a sub network. This mechanism provides an efficient data structure for representing and analyzing hierarchical networks.

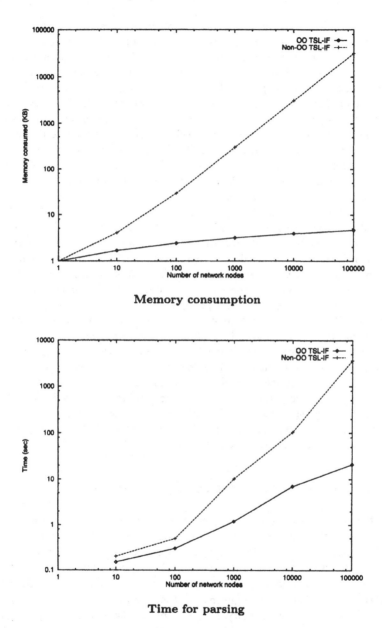

Fig. 2. Quantitative comparison between object-oriented (OO) and non-OO techniques used to construct TSL-IF

Prior to such an OO implementation of the IF, references to sub networks were replaced by their definitions; that is, "elaboration" was performed by cloning (or duplicating) the sub networks every time nested topologies were encountered during parsing. Every node and sub-sub network were recursively cloned, new netlists were constructed, and merged with the outer topology. The elaboration is necessary in order to enable static optimizations and code-generation. This technique provided a simple mechanism to elaborate (or flatten) nested topologies and worked fine for small networks. As the network size increased, the time and memory requirements grew exponentially. In order to circumvent this bottleneck TSL-IF was developed. A quantitative comparison of the the two techniques is shown in Fig. 2. The data was collected on a workstation with dual Pentium II processors (300MHz) with 128MB of main memory running Linux (version 2.1.126). The memory consumption of the parsing routines was monitored by overloading the new and delete calls of C++. As illustrated in the figure, the OO representation dramatically out performed the traditional technique employed earlier. The OO nature of TSL-IF enabled development of a seamless interface with the static analysis and code-generation module, considerably reducing development overheads. The OO representation also enabled "just-in-time" elaboration (or a lazy elaboration) of the sub networks without any change to the other components, thus reducing time and resource consumptions. Analysis of topologies consisting of a million nodes was not feasible with the earlier technique as the system would run out of memory. Failure of the front end to meet the fundamental requirements proved a hurdle for further system development. Using the OO representation efficient analysis of ultra-large topologies, without incurring redevelopment costs, was possible. This is an excellent example to highlight the importance of the OO paradigm. The techniques provided an effective and efficient solution to enable analysis of ultra-large topology specifications. The generation of configuration information in TSL format can be automated to generate different inter-network topologies. Further details on TSL is available in the literature [9].

4.2   Static Analysis and Code-Generation Modules

The static analysis and code-generation modules of USSF play an important role in reducing the memory consumed during simulation. TSL-IF generated by the TSL parser forms the primary input. The static analyzer collates information on the modules repeated in the topology. Using this information, duplicate object descriptions are identified and subsumed to a single entry in the various internal tables. Having a single entry rather than a few hundred thousands dramatically reduces their memory consumption. It also reduces the processing overheads of the data structures during simulation. The code-generator module uses the analyzed intermediate format to generate the final simulation topology. The generated code is compiled and linked with the USSF library, the WARPED library, and the USSF API compliant user developed modules to obtain the final executable. Although the generated code is currently in C++, in compliance with the implementation of WARPED and USSF, the techniques employed are independent of the implementation language.

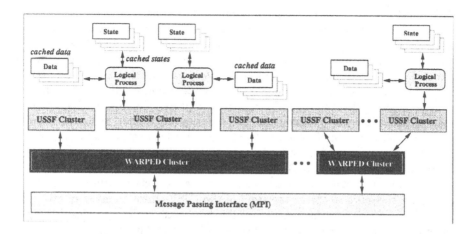

Fig. 3. Layout of an USSF simulation

4.3 USSF Kernel

The core functionality to enable simulation of ultra-large networks is provided by the USSF kernel modules. The kernel is built on top of the WARPED simulator and presents a similar application program interface (API) to the model developer. The core of the USSF kernel is the USSF cluster. The USSF cluster module represents the basic building block of USSF simulations. A USSF cluster performs two important functions; namely (i) it acts as an LP to WARPED; and (ii) it acts as a cluster to the application programmer. As shown in Fig. 3, the USSF cluster is used to group a number of LPs that use the same description together. A single copy of an user process is associated with different data and states to emulate its various instances. In order to enable this feature, the data associated with the different model descriptions need to be decoupled. The USSF API employs standard OO techniques to insulate the model developer from such implementation intricacies. The USSF cluster uses file caches to maintain the data and states of the various processes. The caching helps in regulating the demands on main memory. Separate data and state caches are maintained to satisfy concurrent accesses to data and state spaces and to reduce cache misses. A Least-Recently-Used (LRU) cache replacement policy is used. OO techniques have been used to decouple the various memory management routines from the core. This design not only provides a simple mechanism to substitute various memory management algorithms, but also insulates the USSF cluster from their implementation details.

The USSF cluster is responsible for scheduling the various application processes associated with it. The cluster appropriately translates the calls made by the WARPED kernel into corresponding application process calls. It is also responsible for routing the various events generated by the application to the WARPED kernel. The WARPED kernel permits exchange of events between the

USSF clusters. To enable exchange of events between the various user LPs, the framework cluster translates the source and destination of the various events to and from USSF cluster ids. In order for the USSF kernel to perform these activities, a table containing the necessary information is maintained by the kernel modules. The table is indexed using the unique process ids that need to be associated with each user LP. To reduce the number of entries in this table, a single entry is maintained for a group of LPs sharing a process description. The static analysis phase assigns contiguous ids to processes constructed using the same simulation objects. This fact is exploited to efficiently construct and maintain the table. The framework cluster also maintains a file based state queue in order to recover from rollbacks [3] that can occur in a Time Warp simulation. An simple incremental state saving mechanism with a fixed check-pointing interval is used for this purpose. Since the state of the USSF cluster consists of merely the offsets into this file, the memory overheads of state saving are considerably reduced. A simple garbage collection mechanism triggered by the garbage collection routines in WARPED is used to prune the state queues. Access to the various methods in the framework kernel is provided via a set of simple application program interface (API). The API is illustrated in the following subsection.

4.4   USSF Application Program Interface

The USSF API closely mirrors the WARPED API [3]. This enables existing WARPED applications to exploit the features of the framework with very few modifications. The USSF Kernel presents an API to the application developer for building local processes (LPs) based on Jefferson's original definition [3] of Time Warp. The API has been developed in C++ and the OO features of the language have been exploited to ensure it is simple and yet robust. The API plays a critical role in insulating the model developer from the intricacies involved with enabling ultra-large parallel simulations. The interface has been carefully designed to provide sufficient flexibility to the application developer and enable optimal system performance.

The basic functionality the kernel provides for modeling LPs are methods for sending and receiving events between the LPs and the ability to specify different types of LPs with unique definitions of state. The user needs to derive the LP's description, state and data classes from the corresponding interface classes. The USSF kernel utilizes the interface classes to achieve its functionality such as swapping of data and states, mapping data and state with corresponding LPs, state saving, rollback recovery, and handling interface calls from WARPED. Interfaces for creating and exchanging events are also defined. However, the user is required to override some of the kernel methods. More specifically, the `initialize` method gets called on each LP before the simulation begins. This gives each LP a chance to perform any actions required for initialization. The method `finalize` is called after the simulation has ended. The method `executeProcess` of an LP is called by the USSF kernel whenever the LP has at least one event to process. The kernel calls `allocateState` and `allocateData` when it needs the LP to allocate a state or data on its behalf. Although it is the responsibility of the modeler to assign unique ids to each LP, the static analysis modules in the USSF perform this functionality. The USSF kernel provides all the necessary interfaces needed by WARPED and handles all the overheads involved in a enabling ultra-large simulations providing a simple yet effective modeling environment.

5 Experiments

This section illustrates the various experiments conducted using the ultra-large simulation framework. The network model used to conduct the experiments was a hierarchical network topology constructed by recursively nesting topologies. Since the topologies are elaborated to a flat network, any random network could have been used. In order to ease study and experimentation, a basic topology consisting of ten nodes was recursively used to scale the network models to the required sizes. Validation of the simulations were done by embedding sanity checks at various points in the model. The nodes representing the terminal points in the network generated traffic based on random Poisson distributions. The nodes generated packets of size 64 bytes whose destinations were chosen a normal distribution based on the size of the topology. The models of the network components used in the specification were developed using the framework's API. The TSL parser was used to analyze the hierarchical topologies and generate appropriate network models. The generated models were compiled and linked with USSF and WARPED libraries to obtain the final executable.

Figure 5 presents the time taken for simulating the generated topologies with WARPED and with USSF. Since the APIs of WARPED and USSF are similar, the models were fully portable across the two kernels. The data was collected using eight workstations networked by fast Ethernet. Each workstation consisted of dual Pentium II processors (300MHz) with 128MB of RAM running Linux (version 2.1.126). The simulation time for the smaller configurations is high due to the initial setup costs. An aggressive GVT period was used to ensure rapid garbage collection by WARPED. As illustrated in Fig. 5, for smaller sized models, WARPED performs better than USSF. The reductions in performance is due to the added overheads needed to enable ultra-large simulations. As shown in Fig. 5, the performance of WARPED deteriorates as the size of simulation increases. A study indicated that the drop in performance occurred when the memory consumptions of the parallel processes exceeded the physical memory sizes of the machines and virtual memory overheads began to dominate.

In order to study the scalability issues of the framework, the hierarchical topology consisting of hundred thousand nodes was used. The above mentioned experimental platform was used; and the number of processors used was varied between one and sixteen. Figure 5 presents the timing information obtained from this experiment. The performance of the framework increases as the number of processors are increased, since more computational resources are available for the concurrent processes to use.

As illustrated by the experiments, USSF enables simulation of very large networks, which was the initial goal of the research. The current implementation

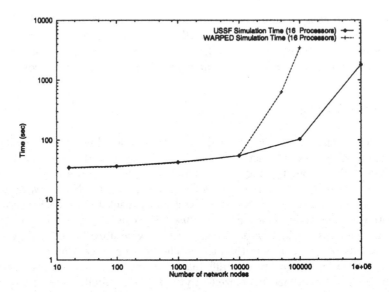

Fig. 4. Simulation time

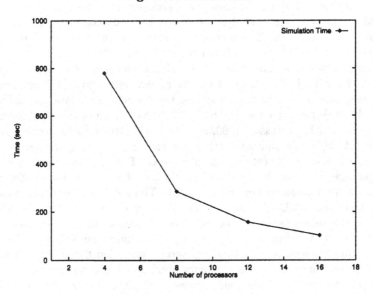

Fig. 5. Scalability data

does not include a number of other proposed partitioning and load balancing optimizations that improve performance of the Time Warp simulations [11]. Although OO design techniques incur some degradation in performance when compared to their non-OO counter parts, the primary motivation to design an OO framework was to develop an simple yet effective simulation environment. Studies are being conducted to improve the performance of USSF.

6   Conclusion

Modern communication networks and their underlying components have grown in size and complexity. Simulation analysis with models built to reflect the ultra-large sizes of today's networks is important in order to study scalability and performance issues. Parallel simulation techniques need to be employed in order to achieve time versus resource tradeoffs. The size and complexity of such parallel simulations requires the system to be carefully developed using standard design techniques such as OO design. Efficient and robust interfaces are necessary to ease application and simulation kernel development. In order to insulate the application developer from such intricacies and to ease modeling and simulation of large networks, an OO framework for simulation of ultra-large networks was developed. The issues involved in the design and development of the framework were presented. The experiments conducted using the framework were illustrated. The need for efficient OO design to enable the system was quantitatively highlighted. From the experimental results the capacity to perform such large simulations in resource restricted platforms was demonstrated. Currently, techniques to improve performance of the framework are being explored. USSF provides a convenient and effective means to model and study ultra-large communication networks of today and tomorrow.

References

1. V. Paxson and S. Floyd. Why we don't know how to simulate the internet. In *Proc. 1997 Winter Simulation Conference*, pp. 44–50, December 1997.
2. A. M. Law and M. G. McComas. Simulation software for communications networks: The state of the art. In *IEEE Communications Magazine*, pp. 44–50, March 1994.
3. R. Radhakrishnan, D. E. Martin, M. Chetlur, D. M. Rao, and P. A. Wilsey. An Object-Oriented Time Warp Simulation Kernel. In *Proc. Int. Symp. Computing in Object-Oriented Parallel Environments (ISCOPE'98)*, volume LNCS 1505, pp. 13–23. Springer-Verlag, December 1998.
4. P. Huang, D. Estrin, and J. Heidemann. Enabling large-scale simulations: Selective abstraction approach to the study of multicast protocols. In *Proc. Int. Symp. Modeling, Analysis and Simulation of Computer and Telecommunication Networks*, October 1998.
5. B. J. Premore and D. M. Nicol. Parallel simulation of TCP/IP using TeD. In *Proc. 1997 Winter Simulation Conference*, pp. 437–443, December 1997.
6. D. M. Rao, N. V. Thondugulam, R. Radhakrishnan, and P. A. Wilsey. Unsynchronized parallel discrete event simulation. In *Proc. 1998 Winter Simulation Conference*, pp. 1563–1570, December 1998.
7. P. A. Wilsey and A. Palaniswamy. Rollback relaxation: A technique for reducing rollback costs in an optimistically synchronized simulation. In *Proc. Int. Conf. on Simulation and Hardware Description Languages*, pp. 143–148. Society for Computer Simulation, January 1994.
8. K. Fall. Network emulation in the Vint/NS simulator. In *Proc. 4th IEEE Symp. Computers and Communications*, July 1999.

9. D. M. Rao, R. Radhakrishnan, and P. A. Wilsey. FWNS: A Framework for Web-based Network Simulation. In *1999 Proc. Int. Conf. Web-Based Modelling & Simulation (WebSim 99)*, pp. 9–14, January 1999.
10. T. J. Parr. *Language Translation Using PCCTS and C++*. Automata Publishing Company, January 1997.
11. R. Fujimoto. Parallel discrete event simulation. *CACM*, 33(10):30–53, October 1990.

Exploiting Parallelism in Real-Time Music and Audio Applications

Amar Chaudhary, Adrian Freed, and David Wessel

University of California, Berkeley, CA, USA

Abstract. We introduce a scalable, extensible object-oriented system developed primarily for signal processing and synthesis for musical and multimedia applications. The main performance issue with these applications concerns functions of discrete-time. Novel techniques exploit fine-grain parallelism in the calculation of these functions to allow users to express them at a high-level in C++. New scheduling strategies are used to exploit symmetric multiprocessors with emphasis on special hard real-time constraints.

1 Introduction

OSW, "Open Sound World," is a scalable, extensible object-oriented language that allows sound designers and musicians to process sound in response to expressive real-time control [1]. OSW is a "dataflow programming language," similar to Ptolemy [2] and Max/MSP [3] in the signal-processing and computer-music communities, respectively. In OSW, components called *transforms* are connected to form dataflow networks called *patches*. OSW is also an "object-oriented" language in which transforms are instances of classes that specify their structure and behavior. OSW allows users to develop at multiple levels including visual patching (as seen in Fig. 1), high-level C++ and scripting. OSW includes a large set of standard transforms for basic event and signal processing. which can be easily extended to include more advanced operations. Since the data types used by transforms are C++ types (i.e., classes or primitive scalars), it is straightforward to add new data types as well.

2 Exploiting Fine-Grain Parallelism in OSW

The recently completed standardization effort for C++ introduced several new features, many of which directly address efficiency issues. Although many of these techniques are being adopted in the numerical-computing community, their use has not been explored in signal-processing and music synthesis applications. Additional techniques exploit mathematical properties of the functions of time used extensively in these applications.

S. Matsuoka et al. (Eds.): ISCOPE'99, LNCS 1732, pp. 49–54, 1999.
© Springer-Verlag Berlin Heidelberg 1999

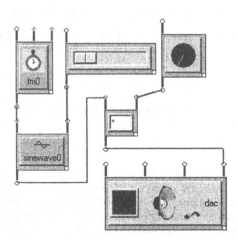

Fig. 1. A patch that plays a pure tone with varying frequency and amplitude

2.1 The Externalizer

OSW includes a graphical tool called the Externalizer that allows users to "peer under the hood" of a transform and extend its behavior without a deep knowledge of C++ or low-level efficiency concerns.

A transform is specified as a collection of inlets, outlets, *state variables* and *activation expressions* that a user can view or modify. A state variable is a public variable of a transform that can be queried or modified by other transforms in OSW. Inlets and outlets are special cases of state variables used in connections. An activation expression is a piece of C++ code that is executed when inlets or state variables are modified. It is specified by the variables that will trigger this activation, whether it should occur immediately or be delayed by a certain amount of time, and the code that should be executed. Consider the following specification of Sinewave, a transform that implements a simple sinusoid oscillator:

Sinewave. Generates a pure tone (i.e., sine wave) signal.

	Name	Type	Default
Inlets	timeIn	Time	
	frequency	float	440.0
Outlets	samplesOut	Samples	
Inherited	SampleRate	float	44100.0
	NumberOfSamples	int	128

Activation Expression activation1, depends on timeIn, no delay
samplesOut = sin(TWOPI * frequency * timeIn);

The activation expression looks like a continuous function of time. However, as we will discuss in the next section, it is actually computing a sequence of samples from a discrete time variable, timeIn. The state variables NumberOfSamples and SampleRate are *inherited* from a more general class of *time-domain transforms* that manipulate time-domain samples.

An Externalizer transform specification is automatically converted to a C++ class which is then compiled into a dynamic library. The new library will by automatically loaded when the transform is first instantiated. Users can also specify new data types, which are converted to C++ struct definitions for use in transforms.

2.2   Optimizations on Functions of Time

Externalizer specifications allow users to specify activation expressions using intuitive, familiar mathematical definitions instead of hand-optimized computer code. This is achieved through the use of function and operator overloading in expressions that use osw::vector<T>[1] OSW's time data type and vector template class. Composition closure, functors and operator overloading are already well-known techniques in the numeric community, and are incorporated into several C++ numeric libraries [4]. We can exploit additional optimizations for functions of discrete time that are used extensively in signal processing applications.

Consider the Time data type used in the Sinewave example. The overloaded sin function in the activation expression expands into a loop that fills a floating-point vector with values of the sine function over a regular sampling interval:

```
Samples temp; float t; int i;
for (t = timeIn->prev_time, i = 0;
     t < timeIn;
     t += timeIn->sampling_interval, ++i ) {
   temp[i] = sin(TWOPI * frequency * t);
}
samplesOut = temp;
```

In this example, prev_time is the previous value assigned to timeIn. The difference between the previous and current values of timeIn, divided by NumberOf-Samples, is sampling_interval. This implementation has the virtue of simplicity, but is too slow for real-time work since the sine function is computed for every sample point.

The usual way to address this deficiency is to replace calls to the mathematical functions with calls to low-level optimized functions. We prefer another approach where we evolve the Time class to include formal descriptions of the mathematical identities behind the optimizations, leaving the compiler to deal with the details of exploiting the identities for the particular machine the code

---

[1] The explicit namespace osw:: is used to avoid confusion with the Standard Template Library class vector<T>.

will run on. The first step is to change the expansion of sin to more honestly reflect the fact that we are really computing discrete-time sequences:

```
for (i = 0; i < NumberOfSamples; ++i) {
   temp[i] = sin(TWOPI * frequency
      * (timeIn->prev_time + i * timeIn->sampling_interval));
}
```

Noting that the sequence $s_0 = 1, \ldots, s_n = s_{n-1}e^k$ computes e^{kn} for real and complex values of k, we can optimize computation of sinusoids using the imaginary result of Euler's identity $e^{i\theta} = \cos\theta + i\sin\theta$. Further noting that k, which is the product of 2π, the frequency and sampling interval, is a loop invariant, it can be calculated once prior to the loop.

```
float factor = exp(complex<float>(0.0,TWOPI * frequency
                   * timeIn->sampling_interval)));
for (i = 0; i < NumberOfSamples; ++i) {
    temp[i] = imag(timeIn->previous_sample * factor);
    timeIn->previous_sample = temp[i];
}
```

The property previous_sample has been added to the Time class to save the previous sample between activations. The inner loop of the expression has been essentially reduced to a single multiply operation via overloading, inlining and standard compiler optimizations. This technique can be extended to the other common trigonometric and exponential functions used in signal processing.

3   Coarse-Grain Parallelism

The real-time scheduler used by OSW supports symmetric multiprocessor computers, as well as configurations with multiple audio devices and time sources. Synchronization primitives (i.e., locks) are included to protect against non-deterministic behavior that arises in such parallel systems without severely compromising performance.

3.1   A Parallel Scheduler for OSW

OSW patches are a special case of dataflow process networks [5]. Music and audio applications exhibit a coarser-grain parallelism in which long chains of transforms must be executed in sequence, but multiple such chains can run in parallel. Examples of such chains include several channels of audio, or synthesizers with multiple voices (i.e., polyphony). If a transform has multiple outlets, the connections to each outlet will start new chains, so any activation expressions triggered by these connections will be added to a queue instead of being executed directly. Given N processors, we instantiate N threads for processing the queued expressions. Each thread executes the following loop *ad infinitum*:

loop
> **pop** an activation expression off the queue and execute it.

end loop

Low-priority tasks such as deallocation of dynamic memory are executed when there are no activations waiting to be scheduled. Higher-priority non-real-time tasks, such as handling user-interface events, are scheduled with a coarser granularity specified in a state variable. Additional threads are needed for handling some asynchronous input devices, such as MIDI input ports. Although this means there will be more threads than processors, the asynchronous input threads will likely be idle much of the time and therefore do not need to be assigned dedicated processors.

3.2   Reactive Real-Time Constraints

OSW is designed for implementing *reactive real-time* audio and music applications. Reactive real-time involves maintaining output quality while minimizing *latency*, the delay between input and output of the system, and *jitter*, the change in latency over time [6]. Because of the combination of human sensitivity for jitter and the need for reactive response to gestures, we have set the latency goal for the OSW scheduler of 10 ± 1ms [7] [8].

It is the job of the audio output device to implement these constraints. The audio output device is a transform that has two state variables that represent these constraints. SampleBufferSize is the number of samples that are sent to the device at once, and TargetLatency is the total number of samples that are allowed to be placed in the output queue awaiting realization by the sound hardware. In order to fulfill real-time requirements, the audio output device has to be able to determine when the signal processing that produces the samples it will output is performed. This is accomplished by controlling *virtual time sources* via the *clock*. Clocks measure real time from hardware devices. Virtual time [9] is a scaleable representation of time. In OSW, virtual time is handled by *time machines*, transforms that scale input from clocks or other time machines. The audio output device includes an activation expression that depends on a clock:

```
while(SamplesInQueue()>TargetLatency-SampleBufferSize){ Wait();}
FlushSamples();
clock = clock + SampleBufferSize / SampleRate;
```

The Wait operation is system dependent, and may include deferring to another thread or process to perform other events such as MIDI input. FlushSamples outputs the samples for this period. The number of samples output is the sample-buffer size.

When the clock is updated, it triggers several activation expressions in a specific order. The clock first triggers the audio input device, which reads a period of samples into a buffer and then activates any audio-input transforms in the program. All the time machines synchronized to this clock are then activated. Finally, the clock re-triggers the activation expression of the audio output device.

Real-time audio contraints require that synchronicity guarantees be added to the process network scheduling described in the previous section. All the transforms that are connected to time-machines synchronized to the same clock must be executed exactly once each clock period. Because OSW allows multiple audio devices and clock sources with different sample rates and periods, it may be necessary to maintain several separate guarantees of synchronicity for each clock source.

4   Discussion

OSW runs on PC's running Windows NT/98 or Linux, and is being ported to SGI workstations. OSW has been used successfully in live musical performances [10]. See http://www.cnmat.berkeley.edu/OSW for more information.

Acknowledgements

We gratefully acknowledge the NSF Graduate Research Fellowship Program for their support of this research. We would also like to thank Lawrence A. Rowe, director the Berkeley Multimedia Research Center, for his support, and Matthew Wright for his contributions to the early design.

References

1. A. Chaudhary, A. Freed, and M. Wright. "An Open Architecture for Real-time Audio Processing Software". *107th AES Convention*, New York, 1999.
2. J. Davis et al. "Heterogeneous Concurrent Modeling and Design in Java". Technical Report UCB/ERL M98/72, EECS, University of California, November 23, 1998. http://ptolemy.eecs.berkeley.edu.
3. D. Zicarelli. "An Extensible Real-Time Signal Processing Environment for Max". *Int. Computer Music Conf.*, pp. 463–466, Ann Arbor, MI, 1998.
4. T. Veldhuizen. "Scientific Computing: C++ Versus Fortran". *Dr. Dobb's Journal*, 22(11):34, 36–8, 91, 1997.
5. E. A. Lee and T. M. Parks. "Dataflow Process Networks". *Proc. IEEE*, 83(5):773–799, 1995.
6. E. Brandt and R. Dannenberg. "Low-Latency Music Software Using Off-the-Shelf Operating Systems". *International Computer Music Conference*, pp. 137–140, Ann Arbor, MI, 1998. ICMA.
7. E. Clarke. "Rhythm and Timing in Music". Diana Deutsch, editor, *The Psychology of Music*, pp. 473–500. Academic Press, San Diego, 1999.
8. M. Tsuzaki and R. D. Patterson. "Jitter Detection: A Brief Review and Some New Experiments". A. Palmer, R. Summerfield, R. Meddis, and A. Rees, editors, *Proc. Symp. on Hearing*, Grantham, UK, 1997.
9. R. Dannenberg. "Real-Time Scheduling and Computer Accompaniment". Max Matthews and John Pierce, editors, *Current Research in Computer Music*. MIT Press, Cambridge, MA, 1989.
10. A. Chaudhary. "Two-Tone Bell Fish", April 25 1999. Live performance at CNMAT/CCRMA Spring 1999 Concert Exchange. Stanford University.

ARAMIS: A Remote Access Medical Imaging System*

David Sarrut and Serge Miguet

Universit Lumière Lyon 2, France

Abstract. In hospital services, practitioners need to access and study large data volume (3D images) with help of specific, parallel, high performance processing tools. This work describes the ARAMIS platform (A Remote Access Medical Imaging System), which allows transparent remote access to parallel image processing libraries. Such system is based on a communication protocol which takes as input parallel libraries (written in C) and leads to objects, which can be combined easily. The end-user application is a Java applet, allowing any common workstation to activate, in a convivial way, time-consuming parallel processing.

1 Introduction

As a part of a project called "Health and HPC" whose goal is to bring High Performance (HP) Computing resources in hospital services, our team focuses on HP image processing tools acting on large data volumes (such as 3D images). In hospital services, 3D images databases are distributed over several linked services (cardiology, radiology), potentially accessible from anywhere through a local network. Such images represent a considerable volume of distributed data which must be accessed and visualized by the practitioners with specific medical image processing tools. For some years, the research community on parallel algorithm for image processing has developed a large number of powerful algorithms and methods, dedicated to MIMD architecture (from fine-grained to mid-grained parallelism, including clusters of workstations). However, these tools are often very optimized and fully efficient at the expense of a limited accessibility, either for end-user practitioners or programmers who want to combine several tools and build large and accessible applications. Moreover, powerful parallel machines are costly and few of them could be present in a same hospital.

We propose a system which allows to easily build (in Java) end-user applications with friendly GUI and with transparent access to remote resources. This system is called ARAMIS (A Remote Access Medical Imaging System). Many modifications has occurred from the beginning of the project [1]. Especially, the whole communication protocol has been fully rewritten in an object-oriented way and it allows to integrate the parallel libraries in a much simpler way. Moreover, the system can now manage dynamically several remote servers. Section 2 refers

* This work is supported by the Région Rhône-Alpes under the grant "*Santé et Calcul Haute-Performance*" (Health and High Performance Computing).

S. Matsuoka et al. (Eds.): ISCOPE'99, LNCS 1732, pp. 55–60, 1999.

to related works and Sect. 3 presents an overview of ARAMIS. We conclude in Sect. 4 with future work.

2   Related Work

Metacomputing techniques allow a same application to access and use various remote resources ([2]). In order to refer our project through the Metacomputing community, we briefly present related projects or general purpose communication schemes which have been already developed.

The NetSolve project [3] aims at bringing remote access to scientific computing libraries. But, even if our goal is also to offer access to remote libraries, the nature of the data (images) and the processing tools are very different. Our approach is specifically oriented to medical image processing and is not suited for other general data types. Moreover, more general projects such as NetSolve or Globus [4] concern Internet-wide computing, whereas in our approach, the set of remote resources is acceded through a *local* network in order to keep efficiency.

[5] described a software architecture, built over the Globus technology [4], which enables from generation to visualization of Computed Microtomography images. The system deals with Gigabytes data volumes and parallel algorithms (reconstruction of microtomographic datasets from numerous slices). However, the visualization part of the project is built over hardware-optimized libraries. As our requirements are to provide accessibility to HP visualization tools from *any* common workstation (via a Web browser), this approach does not seem to be suitable in our case (even if such hardware acceleration could also be used in the libraries standing on the servers).

Because of the heterogeneity of both used languages and involved machines, we think to investigate the use of the CORBA technology in future work. However, such a choice would imply several issues (most of the libraries are not oriented-object, do all the browsers include ORB?) and a large amount of code to be added to the server. Other general framework such as the IceT project [6] or Nexus [7] (the communication module of the Globus project previously cited), are presently under investigation.

3   Overview of ARAMIS

In hospital services, we consider several databases storages (corresponding to different services or images acquisition devices), and a few number of powerful machines (which could be real multi-processor machines, or NOWs: Networks Of Workstations). The parallel machines will act as powerful graphics computing and rendering servers. A server stands also for databases management, and makes use of the parallel server through several high-performance mechanisms [8]. The databases sites and the parallel machines are linked by a high-speed network, independent of the network which links the end-user machines with the computing resources. Hence, this scheme defines two levels of network: the first one (between databases and servers) supports transport of large volume of data, and the second one could be classical low bandwidth network, because it only sends 2D images from parallel machines to practitioner's workstation. Special attention have been paid to make use of existing materials, and to propose cheap hardware configuration.

3.1 Java Access to Remote HP Image Processing Libraries

We manage a set of libraries, each of them consisting in several image processing algorithms. These libraries has been developed for some years (in C language) and with the parallel communication library called PPCM [9], which allows to compile the code for any defined target platforms (PVM, MPI, or parallel architecture such as Cray T3E or Intel Paragon). The range of image processing covered by these tools consists in time-consuming tasks requiring high computing capabilites (both memory and CPU usage). For instance: parallel Volume Rendering [10], parallel (real-time) surface extraction with the Marching-Cubes algorithm [11], parallel Z-Buffer [12], (sequential) optimized 3D volumes registration [13] and so on. According to a set of mechanisms embedding the low-level communication details, we can access to these libraries with a Java applet. The application is thus encompassed in a Web browser or can also be used as a standalone application. It should be noticed that, by this way, the practitioner does not have to change his usual work environment nor his workstation.

3.2 Data Flow

The large 3D volumes of data are *never* transfered to the user's workstation. The applet only receives 2D images, resulting from a remote processing. For example, for a simple visualization purpose, only 2D (compressed) slices of volumes are sent on request (see Fig. 1), when acting on a slider. Such a process is fast enough to provide interactive displaying on an Ethernet local network. Moreover, we consider three different kinds of tasks:

- the low resource-consuming tasks are done on user's machine (colors enhancement for instance, see Colormap Editor in Fig. 1).
- the high resource-consuming tasks are done remotely and using remote data. Typically, it concerns processes which would normally take several minutes to complete on a monoprocessor workstation, and which are accomplished in few seconds with this system.
- however, some *interactive* tasks (real-time 3D visualization of millions of polygons for instance) would only be displayed at a low frame rate due to the network overhead. Thus, instead of overflowing the network with such particular tasks and in order to keep usability for very weak end-user workstation, we advocate the use of a two-steps method. The choice of the spatial position of the 3D object is done locally through a simplified interface, and the full-resolution and true-color rendering is done remotely on user's request (or when mouse releases the virtual track-ball).

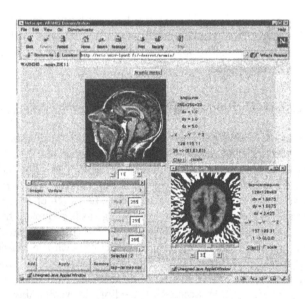

Fig. 1. The Java applet embedding a Slice deliver. The two floating windows area Colormap Editor and another Slice deliver.

This approach allows to spread tasks on remote server or local machine according to its nature. Hence, the server is not bother by a considerable number of different tasks which can be done locally and can thus quickly serve the clients.

3.3   Providing Simple Objects to Control Remote Resources

The underlying protocol is built with classical object-oriented methods, but is adapted for our purpose and particularities with the goal of reducing as most as possible the network overhead. Our system stands between two kind of developers: those who provide HP parallel tools, and those who build end-user applications dedicated to the specific needs of practitioners. The former should depose their libraries on a repository and provide as few as possible supplementary code to integrate their tools into the system. The latter should only manipulate some high-level objects which interact transparently with the remote processes.

In this way, at the lowest-level of the protocol (Transport layer), we use TCP/IP as data transport with the classical socket interface, because of the integration with the C language and the Unix architecture used on most of the parallel machines. At a higher level (Remote Reference layer), we choose to developed a specific purpose *stub-skeleton* scheme:

On server's side, several entry points are extracted from each libraries and are inserted into the system by adding simple functions. This step is presently done manually by the library's developer but require very few lines. Because of the very limited set of data types (3D matrix of voxels [14], or lists of polygons), a true IDL (Interface Definition Language) is not yet provided.

On client's side, simple object (which act as *stub*) are then provided. They embed the communication details and allow to activate and wait (asynchronously with use of Java thread) for remote processing. The data remote references on client's side are also lightweight objects.

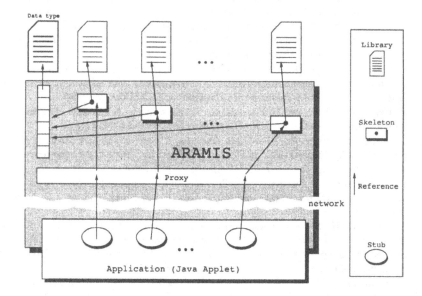

Fig. 2. Overview of ARAMIS. The grey part depicts the hidden protocol which allows to connect Java objects to remote libraries

4 Conclusion

This work deals with a system called ARAMIS (A Remote Access Medical Imaging System) which allows practitioners to activate HP medical images processing with a friendly GUI in their Web browser. The system is built with a communication protocol taking as input parallel libraries (written in C) and leading to Java objects, which can be combined in order to build end-user applications. This project is still in development and integration of all the parallel libraries are in progress. Moreover, as parallel image processing techniques often need to spread data over the different processors for efficient load-balancing, data must be redistributed between two executions of different treatments. This is expected to be done automatically with help of the *ParList* data redistribution algorithm [15], which allows well balanced workload while keeping the overcost of the redistribution at a small value. Hence, before each parallel algorithm execution, a simple and automated call to this library will allows to distribute the data in an efficient way for the process. Such a strategy is currently not yet fully integrated into ARAMIS, but we actively pursuing this goal.

References

1. D. Sarrut. ARAMIS: an "on line" Parallel Platform for Medical Imaging. In H. R. Arabnia, editor, *International Conference on Parallel and Distributed Processing Technique and Applications*, pages 509–516. CSREA Press, July 1998.
2. I. Foster and C. Kesselman. The Globus project: A status report. In *IPPS/SPDP '98 Heterogeneous Computing Workshop*, pages 4–18, 1998.
3. H. Casanova and J. Dongarra. NetSolve: A Network-Enabled Server for Solving Computational Science Problems. *The International Journal of Supercomputer Applications and High Performance Computing*, 11(3):212–223, 1997.
4. K. Czajkowski, I. Foster, N. Karonis, and C. Kesselman. A Resource Management Architecture for Metacomputing Systems. *LNCS* 1459:62–81, 1998.
5. G. von Laszewski et. al. Real-Time Analysis, Visualization, and Steering of Microtomography Experiments at Photon Sources. In *Ninth SIAM Conference on Parallel Processing for Scientific Computing*, April 1999.
6. P.A. Gray and V.S. Sunderam. IceT: Distributed Computing and Java. In *ACM Workshop on Java for Science and Engineering Computation*, volume 9:11, November 1997.
7. I. Foster, G. K. Thiruvathukal, and S. Tuecke. Technologies for Ubiquitous Supercomputing: a Java interface to the Nexus communication system. In G. C. Fox, editor, *Java for Computational Science and Engineering — Simulation and Modeling*, volume 9:6, pages 465–476, December 1997.
8. L. Brunie and E. Mayer. Distributed Systems and Databases. In *4th International Euro-Par Conference, Southampton*, volume 1470. LNCS, September 1998.
9. H.P. Charles, O. Baby, A. Fouilloux, S. Miguet, L. Perroton, Y. Robert, and S. Ubéda. PPCM: A Portable Parallel Communication Module. Technical Report 92-04, Ecole Normale Supérieure de Lyon, 1992.
10. J.J. Li and S. Miguet. Parallel Volume Rendering of Medical Images. In Q. Stout, editor, *EWPC'92: From Theory to Sound Practice*, Barcelona, 1992.
11. S. Miguet and J.M. Nicod. An Optimal Parallel Iso-Surface Extraction Algorithm. In *Fourth International Workshop on Parallel Image Analysis (IWPIA'95)*, pages 65–78, December 1995.
12. H.P. Charles, L. Lefevre, and S. Miguet. An Optimized and Load-Balanced Portable Parallel ZBuffer. In *SPIE Symp. Electronic Imaging: Science and Technology*, 1995.
13. D. Sarrut and S. Miguet. Fast 3D Images Transformations for Registration Procedures. In 10th *International Conference on Image Analysis and Processing*. IEEE Comp. Society Press, September 1999. To appear.
14. S. Miguet. Voxcube: a 3D imaging package. Technical Report 92-05, Ecole Normale Supérieure de Lyon, 1992.
15. F. Feschet, S. Miguet, and L. Perroton. ParList: a Parallel Data Structure for Dynamic Load Balancing. *J. Par. and Distr. Computing*, 51:114–135, 1998.

Language Interoperability for High-Performance Parallel Scientific Components*

Brent Smolinski, Scott Kohn, Noah Elliott, and Nathan Dykman

Lawrence Livermore National Laboratory, Livermore, CA, USA

Abstract. Component technologies offer a promising approach for managing the increasing complexity and interdisciplinary nature of high-performance scientific applications. Language interoperability is required for components written in different languages to communicate. In this paper, we present an approach to language interoperability for high-performance parallel components. Based on Interface Definition Language (IDL) techniques, we have developed a Scientific IDL (SIDL) that focuses on the abstractions and performance requirements of the scientific domain. We are developing a SIDL compiler and the associated run-time support for reference counting, reflection, object management, and basic exception handling. The SIDL approach has been validated for a scientific linear solver library. Initial timing results indicate that the performance overhead is minimal (less than 1%), whereas the savings in development time for interoperable software libraries can be substantial.

1 Introduction

The scientific computing community is beginning to adopt component technologies and associated programming methodologies [1,2,3,4] to manage the complexity of scientific code and facilitate code sharing and reuse. Components require language interoperability to isolate component implementation details from applications. This ensures that applications and components can be created and evolve separately. With the proliferation of languages used for numerical simulation—such as C, C++, Fortran 90, Fortran 77, Java, and Python—the lack of seamless language interoperability negatively impacts the reusability of scientific codes.

Providing interoperability among the many languages used in scientific computing is a difficult problem for both component and library developers. Without language interoperability, application developers must use only the same language as the components, even though better languages may exist. If language interoperability is desired, component developers and users are often forced to write "glue code" that mediates data representations and calling mechanisms between languages. However, this approach is labor-intensive and in many cases

---

* Work performed under the auspices of the U.S. Department of Energy by Lawrence Livermore National Laboratory under Contract W-7405-Eng-48. This work has been funded by LDRD UCRL-JC-134260 grant 99-ERD-078.

S. Matsuoka et al. (Eds.): ISCOPE'99, LNCS 1732, pp. 61–71, 1999.

does not provide seamless language integration across the various calling languages. Both approaches couple the components and applications too tightly, restricting component reuse and flexibility.

1.1 Language Interoperability Design Considerations

The design considerations associated with language interoperability for high-performance scientific computing differ from those of the business sector, which is supported by industry efforts such as COM [5,6] and CORBA [7]. The Common Component Architecture (CCA) [1], Equation Solver Interface [8] and other scientific computing working groups require support for complex numbers, Fortran-style dynamic multidimensional arrays, object-oriented semantics with multiple inheritance and method overriding, and very efficient function invocation for components living in the same address space. The CCA consortium is developing component technologies appropriate for high-performance parallel scientific computing. The ESI is developing standards for linear solvers and associated preconditioners based on component approaches to increase the interoperability of numerical software developed by different development teams.

1.2 Related Interoperability Approaches

Several language interoperability packages have been developed that automatically generate glue code to support calls among a small set of targeted languages. For example, the SWIG package [9] reads C and C++ header files and generates the mediating code that allows these routines to be called from scripting languages such as Python. Such approaches typically introduce an asymmetric relationship between the scripting language and the compiled language. Calls from the scripting language to the compiled language are straight-forward, but calls from the compiled language to the scripting language are difficult or are not supported.

Foreign invocation libraries have been used to manage interoperability among targeted languages. For instance, the *Java Native Interface* [10] defines a set of library routines that enables Java code to interoperate with libraries written in C and C++.

Such interoperability approaches support language interoperability among only a limited set of languages, and they do not support a single, universal mechanism that works with all languages. In the worst case, interoperability among N languages could require $O(N^2)$ different approaches. Component architectures require a more general approach, which we describe in the following section.

1.3 Interoperability Through an IDL Approach

One interoperability mechanism used successfully by the distributed systems and components community [6,7,11,12] is based on the concept of an Interface Definition Language or IDL. The IDL is a new "language" that describes the calling interfaces to software packages written in standard programming languages such as C, Fortran, or Java. Given an IDL description of the interface, IDL compilers automatically generate the glue code necessary to call that software component from other programming languages. The advantage of an IDL approach over other approaches is that it provides a single, uniform mechanism for interoperability among a variety of languages.

Current IDL implementations are not sufficient for specifying interfaces to high-performance scientific components. First, standard IDLs such as those defined by CORBA and COM are targeted towards business objects and do not include basic scientific computing data types such as complex numbers or dynamic multidimensional arrays. Second, approaches focused on distributed objects do not generally provide support for high-performance, same address space function calls between different languages. Our performance goal is to reduce the overhead of single address space function calls to about that of a C++ virtual function invocation. Third, many IDLs do not support multiple inheritance, which is a requirement for some of our customers [8,13,14], or have a limited object model. For example, COM does not support multiple inheritance and supports implementation inheritance only through composition or aggregation, which can be computationally expensive and difficult to implement. CORBA does not support method overriding, which is required for polymorphism.

We have adopted an IDL approach for handling language interoperability in a scientific computing environment. We have developed a Scientific IDL called SIDL [15,16] as well as a run-time environment that implements bindings to SIDL and provides the library support necessary for a scientific component architecture. Currently SIDL supports bindings to C and Fortran 77, although others are under development. These languages were chosen first because the bindings to them posed particular challenges in mapping the object-oriented features of SIDL onto procedural languages. We have begun implementing the C++ bindings, which will not posed these same challenges. Preliminary experiments with a scientific solver library have shown that SIDL is expressive enough for scientific computing and that language interoperability is possible with little measurable run-time overheads.

1.4   Paper Organization

This paper is organized as follows. Section 2 introduces SIDL features that are necessary for high-performance parallel computing. Section 3 describes the bindings of SIDL to C and Fortran 77, as well as the run-time environment, which includes a SIDL compiler and library support. Section 4 details the process of applying the SIDL interoperability approach to a scientific software library and provides parallel performance results for both C and Fortran. Finally, we conclude in Sect. 5 with an analysis of the lessons learned and the identification of future research issues.

2   Scientific Interface Definition Language

A scientific IDL must be sufficiently expressive to represent the abstractions and data types common in scientific computing, such as dynamic multidimensional arrays and complex numbers. Polymorphism—required by some advanced numerical libraries [8]—requires an IDL with an object model that supports multiple inheritance and method overriding. The IDL should also provide robust and efficient cross-language error handling mechanisms.

Unfortunately, no current IDLs support all these capabilities. Most IDLs have been designed for operating systems [5,17] or for distributed client-server computing in the business domain [7,11,12] and not for scientific computing.

The design of our Scientific IDL borrows many ideas from the CORBA IDL [7] and the Java programming language [18]. SIDL supports an object model similar to Java with separate interfaces and classes, scientific data types such as multidimensional arrays, and an error handling mechanism similar to Java and CORBA. SIDL provides reflection capabilities that are similar to Java.

The following sections describe SIDL in more detail. An example of SIDL for a scientific preconditioning solver library is given in Fig. 3 of Sect. 4.

2.1   Scientific Data Types

In addition to standard data types such as *int*, *char*, *bool*, *string*, and *double*, SIDL supports *dcomplex*, *fcomplex*, and *array*. An *fcomplex* is a complex number of type float, and a *dcomplex* is a complex number of type double. A SIDL *array* is a multidimensional array contiguous in memory, similar to the Fortran-style arrays commonly used in scientific computing. The *array* type has both a type, such as *int* or *double*, and a dimension, currently between one through four, inclusive. In comparison, CORBA supports only statically-sized multidimensional arrays and single-dimension sequences, and COM supports only pointer-based, ragged multidimensional arrays.

2.2   SIDL Object Model

SIDL needs to support some type of object model to meet the needs of our customers, particularly the Equation Solver Interface Forum [8], as well as internal customers at Lawrence Livermore Labs [13,14]. However, SIDL includes constructs that allow non-object based mappings of SIDL onto procedural programming languages (e.g. C and Fortran 77).

The SIDL object model is similar to that of the Java programming language. We chose the Java object model for SIDL because it provides a simple model for multiple inheritance. SIDL supports both interfaces and classes. A SIDL class may inherit multiple interfaces but only one class implementation. This approach solves the ambiguity problems associated with multiple implementation inheritance in languages such as C++.

SIDL provides a new set of interface method declarations. These declarations provide optimization opportunities and increase the expressiveness of the

IDL. Like Java, class methods may be declared **abstract, final,** or **static**. An **abstract** method is purely declarative and provides no implementation; an implementation must be provided by a child class. A **final** method is one that cannot be overridden by child classes. The **final** construct enables optimizations in the run-time system that eliminate potential dereferences to an overriding method. As in C++ or Java, **static** methods are associated with a class, not a class instance, and therefore may be invoked without an object. The **static** construct simplifies developing SIDL interfaces to legacy libraries (e.g. **Fortran 77** subroutine libraries) that were written without object-oriented semantics.

2.3   Scoping and Exception Handling

Every class and interface belongs to a particular package scope. Packages in SIDL are similar to namespaces in C++ and packages in Java. The **package** construct is used to create nested SIDL namespaces. Packages help prevent global naming collisions of classes and interfaces that are developed by different code teams.

Component architectures require robust error handling mechanisms that operate across language barriers. We have designed an error reporting mechanism similar to Java. All exceptions in SIDL are objects that inherit from a particular library interface called *Throwable*. Error objects support more complex error reporting than what is possible with simple integer error return codes. Error conditions are indicated through an environment variable that is similar to CORBA.

2.4   Reflection

Reflection is the mechanism through which a description of object methods and method arguments can be determined at run-time. Reflection is an critical capability for component architectures, as it allows applications to discover, query, and execute methods at run-time. This allows applications to create and use components based on run-time information, and to view interface information for dynamically loaded components that is often unavailable at compile-time. These capabilities are required by the CCA, which will layer graphical builder tools on top of a SIDL infrastructure.

The SIDL run-time library will support a reflection mechanism that is based on the design of the Java library classes in java.lang and java.lang.reflect. The SIDL compiler automatically generates reflection information for every interface and class based on its IDL description. The run-time library will support queries on classes and interfaces that allow methods to be discovered and invoked at run-time.

3   Bindings and Implementation

SIDL defines component interfaces in a language-independent manner. For each programming language, we define language mappings that map constructs in

SIDL onto that target language. In this section, we describe the mappings of SIDL to C and Fortran 77, as well as the required library support for the run-time environment. We discuss only the more challenging aspects of the mappings and implementation; a complete specification can be found elsewhere [16].

3.1 Mappings to C and Fortran 77

Because SIDL is based on CORBA IDL, we were able to use the CORBA specification [7] as a guide in mapping many of the SIDL constructs into C. Fortran 77 mappings closely followed the C mappings, with exceptions as described below. The mappings for complex numbers and multidimensional arrays to C and Fortran 77, which are not part of the CORBA IDL, where relatively straightforward.

Mapping SIDL classes and interfaces in C and Fortran 77 presented some interesting challenges, since neither language supports object-oriented features. However, the IDL approach allows object-oriented concepts to be mapped onto non-object-oriented languages. For C, SIDL classes and interfaces are mapped to opaque structure pointers that encapsulate private data members, method invocation tables, and other implementation details. For Fortran 77, classes and interfaces are mapped to integers that are used as handles. The run-time environment manages object information and automatically translates between the Fortran integer representation and the actual object reference. Methods on SIDL objects are invoked using a standard C or Fortran 77 function call with the object reference as the first parameter. Figure 3 of Sect. 4 illustrates these conventions for a scientific linear solver library.

3.2 Implementing the SIDL Run-Time Environment

Much of the effort in developing the SIDL compiler and run-time system was in implementing the object model, namely: virtual function tables, object lookup table for mapping to and from Fortran integer handles, reference counting, dynamic type casting, exception handling mechanism, and reflection capabilities. The run-time library support is implemented in C and the compiler is written in Java. The "glue" code generated from the compiler is in C.

All object support is distributed between the glue code and the run-time library. The glue code contains the implementation of the object mapping, including the virtual function lookup table (similar to a C++ virtual function table), constructors, destructors, and support for dynamic type casting. The run-time library contains support for reference counting, object lookup mechanisms necessary for Fortran objects, and exception handling mechanisms. The reflection capability is supported through both the glue code and the run-time library.

One of the goals of the SIDL run-time environment is to provide extremely fast function calls between components living in the same memory space. For C to C calls, our current implementation requires one table look-up (to support virtual functions) and one additional function call. Calls between C and another language add the overhead of an additional function call, and calls between two non-C languages requires yet another call. These additional function calls are needed to isolate language-specific linker names. Where possible, the SIDL compiler takes advantage of the **static** and **final** qualifiers in SIDL by eliminating a virtual function table lookup to methods for those types.

4   Applying SIDL to a Scientific Library

As a test case, we used the SIDL tools to create new interfaces for a semicoursening mulitigrid (SMG) solver [14], a preconditioner that is part of the *hypre* linear solver library [13]. *hypre* is a library of parallel solvers for large, sparse linear systems being developed at Lawrence Livermore National Laboratory's Center for Applied Scientific Computing. The library currently consists of over 30,000 lines of C code, and it has 94 encapsulated user-interface functions. To test our approach, we created a SIDL interface and and created both C and Fortran 77 library wrappers with SIDL. We ran similar test drivers for the two SIDL generated wrappers and the original C interface already provided by the library, and compared the results from all three runs.

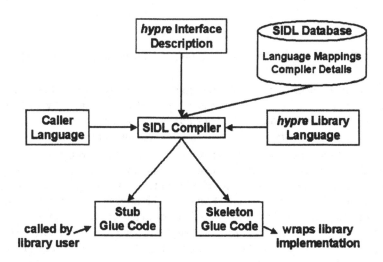

Fig. 1. Generating "glue" code for the *hypre* library using the SIDL tools.

Wrapping *hypre* using SIDL proceeded in three steps. First, the existing *hypre* interface was written in SIDL by two people, one who was familiar with SIDL and another who was familiar with the *hypre* library. The second step was to run the SIDL compiler with the interface description as input to automatically generate the glue code for each class (see Fig. 1). Since the names created by SIDL compiler are slightly different from those expected by the rest of the original *hypre* library, the library had to be slightly modified to match the new names of the SMG interfaces. This step is not required if SIDL conventions are used and

```
package hypre {
  class stencil {
    stencil NewStencil(in int dim, in int size);
    int SetStencilElement(in int index, inout array<int> offset);
  };
  class grid {
    grid NewGrid(in mpi_com com, in int dimension);
    int SetGridExtents(inout array<int> lower, inout array<int> upper);
  };
  class vector {
    vector NewVector(in mpi_com com, in grid g, in stencil s);
    int SetVectorBoxValues(inout array<int> lower,
       inout array<int> upper, inout array<double> values);
    ...
  };
  class matrix { /* matrix member functions omitted in this figure */ };
  class smg_solver {
    int Setup(inout matrix A, inout vector b, inout vector x);
    int Solve(inout matrix A, inout vector b, inout vector x);
    ...
  };
};
```

Fig. 2. Portions of the *hypre* interface specification written in SIDL.

only has to be done once. Once the function calls were manually added for the C language bindings, the Fortran interface was created automatically by running the compiler once more with options for Fortran. The final step was to compile and link the drivers with the skeletons, stubs, and the *hypre* library.

We rewrote an existing SMG test driver to test the performance of the new interfaces. The driver uses SMG to solve Laplace's equation on a 3-D rectangular domain with a 7-point stencil. First, all calls in the existing C driver to the *hypre* library were replaced with the new C interfaces created by SIDL. Then we wrote a new Fortran driver for the same problem that calls the same *hypre* functions via the new Fortran interface. Figure 2 shows a portion of the *hypre* interface written in SIDL, and Fig. 3 shows portions of both the C and Fortran drivers that call the *hypre* library using the automatically generated interfaces.

Both test drivers produced the same numerical results. We compared the efficiency of the new C and Fortran drivers to the original C driver. To set up and solve one problem required 45 calls to functions through wrappers created by SIDL. When used to solve large problems–both sequentially and in parallel on 216 processors–there was no noticeable effect (less than 1%) on the speed of execution, because the overhead of the numerical kernel of the solver far outweighed any function call overhead added by SIDL. The effect of SIDL was more noticeable when using a trivially small problem, which made the same function calls to the solver but required the solver to do much less work. Even so, SIDL added less than 3% to the execution time. This suggests that components

C Test Code

```
hypre_vector b, x;
hypre_matrix A;
hypre_smg_solver solver;
hypre_stencil s;

b = hypre_vector_NewVector(com, grid, s);
...
x = hypre_vector_NewVector(com, grid, s);
...
A = hypre_matrix_NewMatrix(com, grid, s);
...

solver = hypre_smg_solver_new();
hypre_smg_solver_SetMaxItr(solver, 10);
hypre_smg_solver_Solve(solver, &A, &b, &x);
hypre_smg_solver_Finalize(solver);
```

Fortran 77 Test Code

```
integer b, x
integer A
integer solver
integer s

b = hypre_vector_NewVector(com, grid, s)
...
x = hypre_vector_NewVector(com, grid, s)
...
A = hypre_matrix_NewMatrix(com, grid, s)
...

solver = hypre_smg_solver_new()
hypre_smg_solver_SetMaxItr(solver, 10)
hypre_smg_solver_Solve(solver, A, b, x)
hypre_smg_solver_Finalize(solver)
```

Fig. 3. Sample test code calling *hypre* interfaces for C and Fortran 77 generated automatically using the SIDL tools.

in scientific computing should be designed as course grained modules. One would not want to put such an interface within a loop that is executed millions of times.

This entire process required less than an afternoon to generate the SIDL interface, edit the skeleton code, and generate C and Fortran stub code. To put this in perspective, an effort by the *hypre* team to manually generate a Fortran interface for *hypre* required over one person-week of effort. This work was targeted at the Solaris platform. Porting this hand-generated Fortran interface to another platform required a substantial re-write of the interface due to differences in Fortran name representation. Such platform dependencies are managed automatically by the SIDL tools.

5   Lessons Learned and Future Work

We have presented SIDL, a scientific interface definition language, and a run-time that meets the requirements requirements for scientific computing. SIDL borrows heavily from the CORBA IDL and Java programming language, while adding features necessary for scientific computing. SIDL seems to capture the abstractions necessary for scientific computing, as well as new features that a run-time can use to perform optimizations, which are not present in current IDL standards.

The SIDL run-time also provides fast same address space calls, which is important for efficient scientific computation. A comparison using the *hypre* library showed that SIDL added only one to two percent overhead compared to the native interfaces. This is neglible when compared to the great savings in developer costs and flexibility. The SIDL run-time allowed the creation of a Fortran 77 interface in the *hypre* library in a fifth of the time required to create a similar interface by hand.

In the future we will develop bindings for C++, Java, Fortran 90, and Python and implement those bindings. Fortran 90 is challenging since Fortran 90 calling conventions vary widely from compiler to compiler. We will also continue our collaboration efforts with the CCA and ESI working groups. Other ESI specifications will require more expressability from SIDL than the *hypre* interface requires. Features may also need to be added to SIDL to support the specification of high-performance scientific components [1]. For instance, constructs may be needed to specify the semantics of parallel function calls accross components. However, right now it is not clear parallel communication can't be done through interface specifications.

References

1. R. Armstrong, D. Gannon, A. Geist, K. Keahey, S. Kohn, L McInnes, S. Parker, and B. Smolinski. Toward a common component architecture for high performance scientific computing. In *Proceedings the Eighth International Symposium on High Performance Distributed Computing*, 1999.
2. S. Balay, B. Gropp, L. Curfman McInnes, and B. Smith. A microkernel design for component-based numerical software systems. In *Proceedings of the First Workshop on Object Oriented Methods for Inter-operable Scientific and Engineering Computing*, 1998.
3. D. Gannon, R. Bramley, T. Stuckey, J. Villacis, J. Balasubramanian, E. Akman, F. Breg, S. Diwan, and M. Govindaraju. Component architectures for distributed scientific problem solving. *IEEE Computational Science and Engineering*, 1998.
4. S.G. Parker, D.M. Beazley, and C.R. Johnson. *The SCIRun Computational Steering Software System*. E. Arge, A.M. Bruaset, and H.P. Langtangen (Eds.), Modern Software Tools in Scientific Computing, Birkhauser Press, 1997.

5. Guy Eddon and Henry Eddon. *Inside Distributed COM*. Microsoft Press, Redmond, WA, 1998.

6. Microsoft Corporation. *Component Object Model Specification (Version 0.9)*, October 1995. See `http://www.microsoft.com/oledev/olecom/title.html`.

7. Object Management Group. *The Common Object Request Broker: Architecture and Specification*, February 1998. Available at `http://www.omg.org/corba`.

8. Equations Solver Interface Forum. See `http://z.ca.sandia.gov/esi/`.

9. David M. Beazley and Peter S. Lomdahl. Building flexible large-scale scientific computing applications with scripting languages. In *The 8th SIAM Conference on Parallel Processing for Scientific Computing*, Minneapolis, MN, March 1997.

10. JavaSoft. *Java Native Interface Specification*, May 1997.

11. Bill Janssen, Mike Spreitzer, Dan Larner, and Chris Jacobi. *ILU Reference Manual*. Xerox Corporation, November 1997. See
    `ftp://ftp.parc.xerox.com/pub/ilu/ ilu.html`.

12. John Shirley, Wei Hu, and David Magid. *Guide to Writing DCE Applications*. O'Reilly & Associates, Inc., Sebastopol, CA, 1994.

13. E. Chow, A.J. Cleary, and R.D. Falgout. Design of the hypre preconditioner library. In *Proceedings of the First Workshop on Object Oriented Methods for Inter-operable Scientific and Engineering Computing*, 1998.

14. P.N. Brown, R.D. Falgout, and J.E. Jones. Semicoarsening multigrid on distributed memory machines. In *SIAM Journal on Scientific Computing special issue on the Fifth Copper Mountain Conference on Iterative Methods*, 1999.

15. Andy Cleary, Scott Kohn, Steve Smith, and Brent Smolinski. Language interoperability mechanisms for high-performance applications. In *Proceedings of the First Workshop on Object Oriented Methods for Inter-operable Scientific and Engineering Computing*, 1998.

16. S. Kohn and B. Smolinski. Component interoperability architecture: A proposal to the common component architecture forum. In preparation.

17. Eric Eide, Jay Lepreau, and James L. Simister. Flexible and optimized IDL compilation for distributed applications. In *Proceedings of the Fourth Workshop on Languages, Compilers, and Run-time Systems for Scalable Computers*, 1998.

18. James Gosling and Ken Arnold. *The Java Programming Language*. Addison-Wesley Publishing Company, Inc., Menlo Park, CA, 1996.

A Framework for Object-Oriented Metacomputing

Nenad Stankovic and Kang Zhang

Macquarie University, Sydney, NSW Australia

Abstract. Visper is a network-based, visual software-engineering environment for parallel processing. It is completely implemented in Java and supports the message-passing model. Java offers the basic platform independent services needed to integrate heterogeneous hardware into a seamless computational resource. Easy installation, participation and flexibility are seen as the key properties when using the system. Visper is designed to make use of Java features to implement important services like efficient creation of processes on remote hosts, inter-process communication, security, checkpointing and object migration. We believe the approach taken simplifies the development and testing of parallel programs by enabling a modular, object-oriented technique based on our extensions to the Java API, without modifying the language. Our experimental results show that the environment is suitable for course grained parallel applications.

1   Introduction

Wide and local area networks represent an important computing resource for parallel processing community. The processing power of such environments is deemed huge, but they are often made up of heterogeneous hardware and software. A number of tools and libraries have been designed to foster the use of networks to run parallel programs. In the domain of the message-passing model products like the MPI [4] and the PVM [3] transform heterogeneous networks of workstations and supercomputers into a scalable virtual parallel computer. Their aim is to provide a standard and portable programming interface for several high level languages. They have been adapted to a variety of architectures and operating systems. The problem with both systems is that programmers have to build a different executable for each target architecture or operating system and programs must be started manually. The insufficient security and dependence on a shared file system limits their use for large or widely distributed applications.

In this paper we focus on describing a novel metacomputing environment named Visper. The aim of the project is to allow a programmer to make an efficient use of computer networks. Visper is implemented in and for the Java language that, due to its platform independence and uniform interface to system services simplifies the implementation of MIMD parallel applications and the system software needed to support them. Visper is conceived as an integrated metacomputing environment that offers the services to design, develop, test and run parallel programs. It provides an

S. Matsuoka et al. (Eds.): ISCOPE'99, LNCS 1732, pp.72 -77, 1999.

MPI like communication library, and features to spawn and control lightweight processes (i.e. threads) on remote hosts from within the environment. In this paper we will expand upon the mentioned points. Section 2 describes the organization and the main services supported by the environment. Section 3 presents the performance data. The paper concludes with Section 4.

2 Visper

Visper is an interactive, object-oriented environment with tools for construction, testing and execution of SPMD applications, implemented in Java. It is conceived as an environment for research into parallel and distributed Java programming, and as an efficient, reliable and platform independent environment that produces useful results. It can be used by multiple users and run multiple programs concurrently. **Fig. 1** shows the tool organization and the main components. Visper consists of a front-end and a backend that are built on top of two communication techniques: a point-to-point (TCP/IP), and an intranet multicast iBus [8]. The purpose of the front-end is to implement the graphical user interface and to encapsulate as much of the syntax of the model as possible. The backend implements the semantics of the model independent of the front-end design. The overall system operation can be described in terms of the two components.

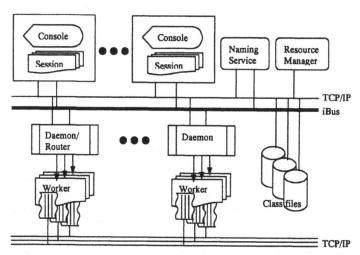

Fig. 1. Overview

Each user starts a console that represents the front-end to the system. The front-end is the only tangible component of the tool. The purpose of the front-end is to enable user-to-system interaction and to implement the syntax of the programming model or program specification. Its main strength is in the visualization of a parallel program composition and execution. Thus, it is the console for the system in which the user constructs a parallel program, then exercises it through an interface. **Fig. 2** shows various components of the console. For example, the front-end provides a visual programming environment with a pallet of model primitives that can be placed on a canvas and which can be interconnected using arcs to create a process communication graph (PCG) [9]. The front-end also generates a structured internal representation of the model, performs syntactic analysis of the graph as the programmer is constructing it, and translates the graph into a Java program. Finally, the front-end must be the user interface to the backend that allows program execution, control and debugging.

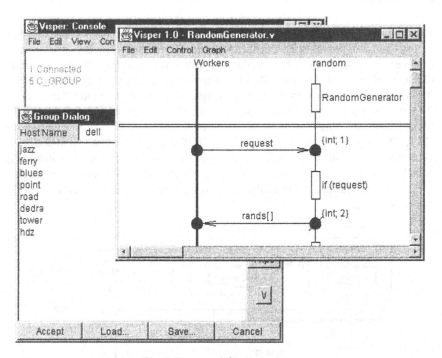

Fig. 2. Front-end Components

Each user can create multiple sessions, where each session can run one parallel program at a time. A session is an ordered collection of processes and represents a virtual parallel computer. It allows control and data collection of the running process. It can dynamically grow or shrink while a program is running, since machines can join or leave (i.e. crash). To minimize the start-up cost, a machine that dynamically joins a session, remains part of it until manually removed by the user. Consequently, the loaded (and JIT compiled) bytecodes are preserved across multiple runs of a program, which also compensates for the code generation overhead that occurs during program execution. (The current generations of JIT compilers do not, however, save the native code for future invocations of the same program [5].) For a HotSpot [12] compiler that also means more time to perform optimizations, as multiple runs are concatenated together.

The backend consists of the services to run parallel programs and generate debugging data. The naming service represents a distributed database of available resources, active sessions, acts as a detector of faulty machines, and as a port-mapper for the point-to-point mode. Each machine that is part of a Visper environment must run a Visper daemon. The daemons fork one worker-process (i.e. one Java Virtual Machine or JVM that runs parallel programs) per session, and maintain workers state (e.g. active, dead). As shown in **Fig. 1**, workers can run multiple *remote threads*. The user can define multiple loading points, and different access modes (e.g. file://, http://) when loading dynamically class files.

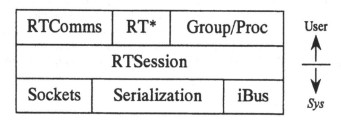

Fig. 3. API

From the programmer's perspective, Visper provides three classes of objects that enable communication and parallel processing (**Fig. 3**) [10]. The RTComms is an MPI like library that provides point-to-point, collective, synchronous and asynchronous communication and synchronization primitives. The RTRunnable interface allows any Java class to be turned into a network-runnable and serializable object (i.e. remote thread). The group and process library provides the classes to group remote threads, create and control them from within a program, and to provide communication scope. Groups are the smallest unit of organization in Visper since, at programer's level, a group creation represents an allocation of resources. Groups can be dynamic and static. A resource manager defines a dynamic group at runtime, while a static group is defined at compile-time and therefore does not scale, but is faster to create. The session layer contains the information about the current session status (e.g. migrated, restarted) and configuration. It provides methods to migrate, restart and checkpoint remote threads. All these services are built on top of the Java Networking API, the iBus middleware and the Java Serialization mechanism [11].

3 Performance

Table 1 summarizes the hardware we have used in our tests. It consisted of 2 Sun boxes running Solaris 2.5 and Sun JDK1.1.6 (no JIT), 2 HP A9000/780/HP-UX B.10.20 with HP-UX Java C.01.15.05, 2 PCs running NT 4.00.1381 and Sun JDK/JRE 1.2. The computers were connected with a 10 Mbps Ethernet.

The results of the COMMS1 (i.e. pingpong) test [2] are summarized as costs in time for messages from 1 to 1000000 bytes. The JavaMPI [6] results were obtained on the LAM system [7] built by gcc2.7.2 on Sun. The test program for iBus is based on the push model that uses different communication channels for sending and receiving messages. This follows the design that is found in RTComms.

Table 1. Hardware

Vendor	Architecture	RAM(MB)	CPU(MHz)
Sun	Ultra 2	256	168 * 2
HP	A 9000/780	512	180
Compaq	Pentium II	64	200
Micron	Pentium II	256	400

Fig. 4 shows the latency vales for the libraries under consideration. Sending small messages in Java is costly, as the best result is an order of magnitude inferior to LAM. However, the time-per-byte cost speaks in favor of Java, since for large messages the overhead caused by Java was offset by the networking cost. The results are summarized in **Fig. 5**. The values for both figures were calculated by linear regression.

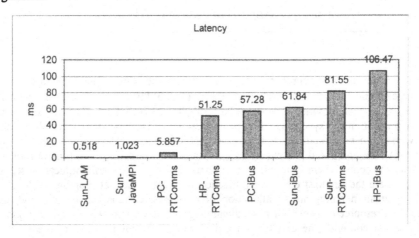

Fig. 4 Latency (ms)

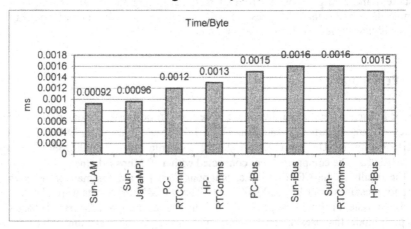

Fig. 5 Time/Byte (ms)

4   Conclusion

Visper is designed as a multi-layer system, where system services are cleanly decoupled from the message-passing API. It has been implemented as a collection of Java classes without any modifications to the language and therefore can be executed on any standard Java virtual machine. It provides secure, object-oriented, peer-to-peer message-passing environment in which programmers can compose, test and run parallel programs on persistent sessions. The RTComms communication class follows the primitives of the MPI standard. Visper supports synchronous and asynchronous modes of execution and primitives to control processes and groups of processes as if they were real parallel computers.

As a future enhancement to the system we can mention better strategy for dynamic resource allocation. The current resource management is rather naive, since it does not take any relevant parameters into account (e.g. computational power, network bandwidth, load, or task size). On the other hand, Java does not provide an interface to the operating system that would allow access to system's statistics, so a solution like the one found in DOME [1] is worth looking at.

References

1. Beguelin, A., Arabe, J. N. C., Lowekamp, B., Seligman, E., Starkey, M., and Stephan, P. *Dome User's Guide Version 1.0*. School of Computer Science, Carnegie Mellon University, Pittsburgh, PA, May 20, 1996.
2. Dongarra, J. J., Meuer, H-W., and Strohmaier, E. The 1994 TOP500 Report. http://www.top500.org/.
3. Geist, G. A., Beguelin, A., Dongarra, J. J., Jiang, W., Manchek, R., and Sunderam, V. PVM 3 User's Guide and Reference Manual. Technical Report ORNL/TM-12187, Oak Ridge National Laboratory, 1993.
4. Gropp, W., Lusk, E., and Skjellum, A. Using MPI, Portable Parallel Programming with the Message-Passing Interface. The MIT Press, 1994.
5. Hsieh, C-H. A., Conte, M. T., Johnson, T. L., Gyllenhaal, J. C., and Hwu, W-M. W. A Study of the Cache and Branche Performance Issues with Running Java and Current Hardware Platforms. *Proceedings of IEEE CompCon'97*, San Jose, CA, 1997, p.211-216.
6. JavaMPI: a Java Binding for MPI. http://perun.hscs.wmin.ac.uk/.
7. Ohio LAM 6.1. MPI Primer / Developing with LAM, 1996. http://www.mpi.nd.edu/lam.
8. SoftWired AG. Programmer's Manual. Version 0.5. August, 20, 1998. http://www.softwired.ch/ibus.htm.
9. Stankovic, N., and Zhang, K. Graphical Composition and Visualization of Message-Passing Programs. SoftVis'97, December, 1997, Flinders University, Adelaide, South Australia, pp.35-40.
10. Stankovic, N., and Zhang, K. Object-Oriented Metacomputing. APPT'99, The Third International Workshop on Advanced Parallel Processing Technologies, Changsha, P. R. China, October 19-20, 1999. http://www.njtu.edu.cn/APPT'99.
11. Sun Microsystems, Inc. Java Object Serialization Specification. Revision 1.4, July 3, 1997, http://java.sun.com.
12. Sun Microsystems, Inc. The Java HOTSPOT Performance Engine Architecture. White Paper, April 1999.

Tiger: Toward Object-Oriented Distributed and Parallel Programming in Global Environment*

Youn-Hee Han[1], Chan Yeol Park[1], Chong-Sun Hwang, and Young-Sik Jeong[2]

[1] Korea University
[2] WonKwang University, Republic of Korea

Abstract. None of the current attempts to provide an Internet-wide global computing infrastructure presents well-defined programming constructs such as object distribution, dispatching, migration and concurrency with maximum portability and high transparency to a programmer. We propose a Web-based global computing infrastructure called *Tiger*, providing well-defined object-oriented programming constructs They allow a programmer to develop a well-composed, object-oriented distributed and parallel application using globally extended resources. We show the performance enhancement by conducting an experiment with a genetic-neuro-fuzzy algorithms.

1 Introduction

The Web has become the largest virtual system, but with most nodes idle. It is appealing to use these hosts for running applications needing substantial computation. Such a computing paradigm has been called *Global Computing*[1]. However, some of the obstacles common to global computing are the heterogeneity of the participating hosts, difficulties in administering distributed applications, and security concerns. The Java language and applets with Java-capable Web browsers have successfully addressed some of these problems. Platform-independent portability is supported by the Java execution environment. Java and applets with Java-capable Web browsers have become a good candidate for constructing a global computing platform[2,3].

However, developing a distributed and parallel application using global resources requires special knowledge beyond that needed to develop an application for a single machine[4,5]. A global system must allow the incremental growth of a system without the user's awareness. For a global system to use object-oriented paradigm, we argue that dispatching and moving an existing object from local machine or a machine to another machine. Both CORBA and JavaRMI, two good solutions constructing distributed applications, do not provide any solution to global computing. Also, in Java, threads are only a mechanism to express concurrency on a single host but do not allow to express the concurrency between

---

* This work is supported in part by Ministry of Information and Communication.

S. Matsuoka et al. (Eds.): ISCOPE'99, LNCS 1732, pp. 78–83, 1999.

remote hosts. Therefore, a gap exists between multithreaded and distributed or global application.

Our challenges are to provide a global distributed and parallel computing infrastructure called *Tiger*, with the programming constructs which combine an existing sequential program with several global application-related properties: object distribution, dispatching, migration and concurrency. *Tiger* provides the programmer with maximum portability and high transparency since it allows one to write programs in a shared-memory style.

2   Tiger System and Object Models

Our system model consists of six kinds of major components: users, brokers, hosting applets, gateways, regions and a manager(Fig. 1).

- *Users* use extra computing power in order to run distributed and parallel applications.
- *Brokers* manage user applications locally, and coordinate communication among the applications and *Tiger*. A user invokes a broker before executing an application.
- *Hosting applets* allow their CPU resources to be used by other users,in the form of Java applets.
- *Gateways* manage parts of hosting applets and coordinate communication among hosting applets and other components in *Tiger*. Each gateway serves exactly one region.
- *Regions* consist of a gateway and hosting applets managed by the gateway. Regions are generated by grouping hosting applets into similar round-trip communication times to a gateway managing them.
- A *manager* registers and manages participating brokers and gateways.

There are two important reasons why our model must provide gateways. The first reason is to distribute manager's massive load. Because of the heavy network traffic generated by many brokers and hosts, the manager may become a bottleneck. To reduce the traffic on the manager, one natural solution is to distribute the manager functions into several other things. The second reason comes from severe limitations on the capabilities of Java applets, since the applets are used to execute untrusted programs from the Internet. Applets cannot create a server socket to accept any incoming connection and Java-capable browsers disallow applets from establishing a network connection except to the machine where they were loaded from. Using gateways as the intermediate message-exchange nodes, we make an applet communicate with any applets within the same region or in different regions. Both broker and gateway intermediate message exchanges between user applications and hosting applets. Also, two different user applications can communicate through each associated broker. *Tiger* is designed as an object-oriented global system which expresses computation as a set of autonomous communicating entities, *objects*. We use a triple-object model in *Tiger*. There are *normal(local) objects*, *distributed objects* and *mobile objects*. Normal

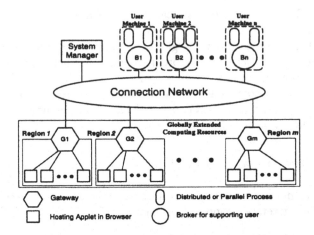

Fig. 1. Tiger system architecture

objects are the same as standard Java objects. Distributed objects are placed to other user machines rather than the one which local objects are placed in. Like distributed objects in CORBA or JavaRMI, they are accessible remotely and locations of them are fixed to the user machine where they are created. Mobile objects are similar to distributed objects except that they can change their current location from a user machine or a hosting applet to other hosting applets. Distributed object and mobile object are the basic units for distribution and concurrency. Users have to specify explicitly which objects are distributed or migrated.

3   Distributed and Parallel Programming

Our programming model concentrates on a clear separation between high-level design and lower-level implementation issues such as object distribution, migration, and control of concurrent activities. When dealing with globally distributed objects, the power of *Tiger* lies in that any client processes or threads in local machine can directly interact with a server object that lives on a globally distributed machine, a user machine or hosting applet through remote method call even when the server object migrates among them.

Tiger provides two application programming interfaces (APIs) related with distributing server objects.

– `turnDistributedObj(TigerObject obj, String name)`

This converts an existing local object, namely `obj`, into a distributed object at any time after its creation. The distributed object becomes accessible remotely and its location is fixed(Fig. 2(a)). A related mechanism is similar to that in the CORBA and JavaRMI.

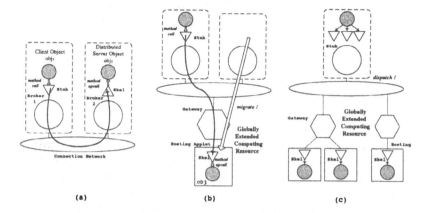

Fig. 2. Objects-distributing and parallel objects-dispatching mechanisms

- `turnMobileObj(TigerObject obj, String name)`
- `turnMobileObj(TigerObject obj, String name, Region destRegion)`

Unlike CORBA and JavaRMI, *Tiger* allows the local or distributed server object, `obj`, to migrate from the creating machine to a hosting applet in globally extended computing resource. Programmers do not present such a information, or they can present only region information, namely `destRegion`(Fig. 2(b)).

On the other hand, *Tiger* provides us with the two API related with dispatching parallel *Tiger*Objects. Parallel programs are largely divided into two parts: one being *the main control program*, which provides a whole body for solving a given problem; the other being *TigerObject*, which describes tasks to be executed in parallel. The main control program must dispatch a number of TigerObjects into the globally extended computing resource and call remote methods on the dispatched objects. Besides, it can watch loads imposed on each region, balance those using object migration. *Tiger* provides us with the two APIs related to dispatching parallel TigerObjects.

- `turnMobileObj(TigerObject obj)`
- `turnMobileObj(TigerObject obj, Region destRegion)`

These overloaded methods dispatch an existing local object, namely `obj`, into a hosting applet in the globally extended resource. The dispatched object becomes accessible remotely from the main control program(Fig. 2(c)). It is not required to provide a name of the object since it is only used by the main control program. *Tiger* itself allocates the globally unique identifier to the object. It is noted that its location is not fixed to the dispatched hosting applet and it can re-migrate to other hosting applets.

Concurrency can be used to perform given applications in parallel on several machine or on one machine controlling different threads, only simulating parallelism. Although method calls of Java are only synchronous, there are three constructs provided to call a remote method in *Tiger*: *synchronous, asynchronous, one-way*. Asynchronous remote method call has a *future* interaction style, in which the caller may proceed until the result is needed. At the time, the caller is blocked until the result becomes available. If the result has been supplied, the caller resumes and continues.

4 Experimental Results

In the experiments, the target application is a parallel genetic-neuro-fuzzy algorithm. These are a hybrid method based on genetic algorithms in order to find the global solution for the parameters of neuro-fuzzy system. They begin by generating an initial population randomly, after encoding the parameters into chromosomes. They run iteratively, repeating the following processes until they arrive at the predetermined ending conditions: *extracting fuzzy rules, self-tuning, fitness evaluation, reproduction, performing genetic operators(crossover and mutation)*. Much computational time is needed to construct a fuzzy system from a chromosome. But the communication time does not affect total processing time. So *Tiger* is suitable to execute such algorithms.

A characteristic of our algorithm is that capability-based adaptive load balancing is supported to reduce total working time in obtaining the optimal system. Let T_i be the time that is taken to execute the operations of chromosmes allocated to Region i and NF_i the sum of the number of fuzzy rules processed in each chromosome allocated to the Region i. Then the number of chromosomes which will be allocated to Region i at next generation, N_i, is defined by

$$N_i = N_c \cdot \frac{C_i}{\sum_{i=0}^{N_f-1} C_i} , \quad where \ C_i = \frac{NF_i}{T_i} \tag{1}$$

N_c is the total number of chromosomes in the system, N_f is the number of regions currently in the system, and C_i is the capability of regions i based on the number of fuzzy rules processed in a unit time.

Using equation (1), we can determine the number of chromosomes which are allocated to each region at the next generation and can move some chromosomes to other regions using *Tiger*'s APIs. The goal of our algorithm is to make a fuzzy system which can approximate the three input nonlinear function defined by $output = (1 + x^{0.5} + y^{-1} + z^{-1.5})^2$.

216 training data were sampled uniformly from input ranges $[1.6] \times [1.6] \times [1.6]$. The system manager and three gateways ran on Pentium PCs with JDK 1.2.1. Ten hosting applets ran on Netscape Communicator 4.5 or Internet Explorer 4.0 on heterogeneous machines connected 10Mb/s Ethernet. Figure 3 shows a positive, but not linear, speedup by the number of hosting applets.

Figure 4 shows the efficiency of the capability-based adaptive load balancing scheme. *NOLB* is no load balancing and *CALB* is capability-based load balancing. When the number of hosting applets is eight, *CALB* shows 1.89 times the performance of *NOLB*. When the number of hosting applets is 10, *CALB* shows 1.80 times the performance of *NOLB*.

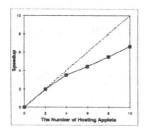

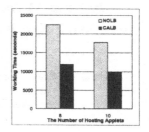

Fig. 3. Speedup for the number of hosting applets

Fig. 4. The efficiency of load balancing

5 Conclusions

We have designed and implemented *Tiger*, a global computing infrastructure for using the computing resources of many machines connected in Web. *Tiger* provides the constructs for object distribution, dispatching, migration and concurrency. Together with the *Tiger* infrastructure, they allow a programmer to develop a well-composed, object-oriented distributed and parallel application using globally extended resources.

References

1. J.E. Baldeshwieler, R.D. Blumofe, and E.A. Brewer, ATLAS: An infrastructure for global computing, *Proc. 7th ACM SIGOPS European Workshop on System support for World Wide Applications*, 1996.
2. K.M. Chandy, B. Dimitrov, H. Le, J. Mandleson, M. Richardson, A. Rifkin, P.A.G. Sivilotti, W. Tanaka, and L. Weisman, A world-wide distributed system using Java and the Internet, *Proc. 5th IEEE Int. Symp. on High Performance Distributed Computing*, Syracuse, NY, August 1996.
3. A. Baratloo, M. Karaul, H. Karl, and Z.M. Kedem, An infrastructure for network computing with Java applets, *ACM Workshop on Java for High-Performance Network Computing*, Palo Alto, California, February, 1998.
4. D. Caromel, W. Klauser, J. Vayssiere, Towards seamless computing and metacomputing in Java, *Concurrency P & E*, pp. 1043-1061, September 1998.
5. M. Boger, F. Wienberg, W. Lamersdorf, Dejay: Unifying concurrency and distribution to achieve a distributed Java, *TOOLS99*, Nancy, France, June, 1999.

SIFFEA: Scalable Integrated Framework for Finite Element Analysis

Xiangmin Jiao, Xiang-Yang Li, and Xiaosong Ma

University of Illinois, Urbana, IL, USA

Abstract. SIFFEA is an automated system for parallel finite element method (PFEM) with unstructured meshes on distributed memory machines. It synthesizes mesh generator, mesh partitioner, linear system assembler and solver, and adaptive mesh refiner. SIFFEA is an implicit parallel environment: The user need only specify the application model in serial semantics; all internal communications are transparent to the user. SIFFEA is designed based on the object-oriented philosophy, which enables easy extensibility, and a clear and simple user interface for PFEM.

1 Introduction

The parallel finite element method (PFEM) is widely used in the computational solution of system simulations in various engineering problems. Generally, there are six steps in the finite element analysis: mathematical modeling, geometric modeling, mesh generation, linear system formulation, numerical solution, and adaptive refinement. Many software packages are available on parallel machines for each individual component of FEM. However, there is a lack of automated integrated systems for PFEM, and a user must do a lot of application coding in integrating softwares from various sources.

To address this problem, we are developing SIFFEA, an integrated framework for FEM on scalable distributed memory machines. SIFFEA includes a mesh generator, a mesh distributor, a linear system assembler, an adaptive mesh refiner, and interfaces with mesh partitioners and linear system solvers. It also contains a novel user interface for specifying the mathematical model. SIFFEA is designed based on object-oriented philosophy, and written in C++. The communications are carried out using MPI for portability. Each component of SIFFEA is highly encapsulated, and the components interact with each other through well-defined interfaces, so that the implementation of one component can be changed without affecting the others.

Below we give a brief description of several classes in SIFFEA: Mesh, MeshGenerator, MeshPartitioner, Solver, and MeshRefiner. A Mesh object encapsulates all information about the mesh, including topological and geometrical data. A MeshGenerator object is activated by a Mesh object. It contains all functions related to mesh generation and will generate triangulations and fill them into the Mesh object. A MeshPartitioner object contains the interfaces with mesh partitioning packages, and is activated by a Mesh object. A Solver object assembles

S. Matsuoka et al. (Eds.): ISCOPE'99, LNCS 1732, pp. 84–95, 1999.

and solves a linear system, and performs error analysis. A MeshRefiner object takes care of adaptive mesh refinement based on the results of error analysis.

Figure 3 is the interaction diagram of the system. It shows the timeline of objects' active periods and operations. The whole timeline can be roughly divided into four parts, shown as S1, S2, S3 and S4 at the left side of the diagram. They correspond to the four stages of the computation: mesh generation, mesh partitioning, linear system assembly and solution, and adaptive mesh refinement. Note that the lengths of objects shown in the diagram are not proportionally scaled to their real life time. The operations in the dashed-line rectangle, i.e., operations in S2, S3 and S4 are repeated until error requirements are satisfied. Notice that this diagram only provides a sequential view of the computations. In fact, except for MeshGenerator and MeshRefiner, each object exists on all the processors and its operations are carried out in parallel.

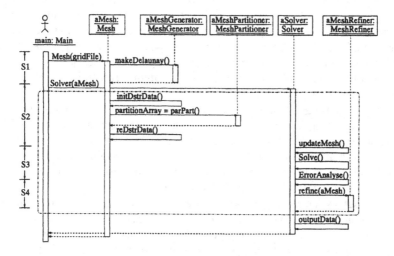

Fig. 1. Interaction diagram of SIFFEA.

SIFFEA provides a clear and simplified user interface. Up to date, the design of SIFFEA has been focusing on linear elliptic problems. For this kind of problems, the user only needs to provide a geometry specification, and a class encapsulating the serial subroutines for computing the element matrices and vectors, and the boundary conditions. All internal communications of SIFFEA are transparent to the user. We are looking into extending SIFFEA for time-dependent and nonlinear problems, while preserving a simple user interface.

The rest of the paper is organized as follows. Section 2 introduces the basic concepts about mesh generation and refinement, and their sequential and parallel algorithms. Section 3 describes the interface with the mesh partitioner, and the algorithms for distributing the mesh onto processors after partitioning. Section 4 addresses the issues about user mathematical model specification and parallel linear system assembly. Section 5 discusses the basic concepts behind the error estimator and the adaptive mesh refiner. Section 6 presents some preliminary experimental results. Section 7 compares SIFFEA with some other PFEM systems. Section 8 concludes the paper with the discussion of future implementation and research directions.

2   Mesh Generation and Delaunay Refinement

Mesh generation is a fundamental problem in FEM. In particular, the *unstructured mesh* is widely used in FEM solvers, due to its advantages of varying local topology and spacing in reducing the problem size, and adapting to complex geometries and rapid changing solutions.

It is well-known that the numerical errors in FEM depend on the quality of the mesh. In particular, it is desirable that the angles of each element are not too small [1,2]. The *Delaunay triangulation* and *Delaunay refinement* algorithms generate high quality meshes satisfying this criterion. Specifically, Delaunay triangulation maximizes the minimum angle among all elements [3]; Delaunay refinement allows one to add extra vertices, called *Steiner points*, in order to further improve the quality of the mesh [4].

SIFFEA contains a two-dimensional mesh generator, which can generate exact Delaunay triangulations, constrained Delaunay triangulations, and conforming Delaunay triangulations on complex domains. The conforming Delaunay triangulations guarantee no small angles, and are thus suitable for finite element analysis. The mesh generator also supports adaptive mesh refinement based on *a posterior* error estimation, which will be discussed in Sec. 5.

The input to the mesh generator is a PSLG [5] description of the geometric domain, which can contain points and line segments. These line segments are called the *constrained edges*. The input domain can also contain holes, as shown in Fig. 2. A clockwise or counterclockwise sequence of edges of a polygon defines a hole. In particular, a clockwise sequence defines a hole inside the polygon, and a counterclockwise sequence defines a hole outside the polygon. Every hole has labels, which specify whether the hole belongs to the mesh, and the element size requirement inside the hole. Boundary conditions may be associated with the constrained edges. To achieve this goal, the user labels each constrained edge with an integer, which will be inherited by the Steiner points on it; then he/she defines a subroutine for computing the boundary condition based on the coordinates and the mark of a point. By convention, the user marks the uninterested edges with 0.

The mesh generation is designed using the object-oriented technology. We separate the mesh representation from the algorithms. Internally, the Mesh class uses the efficient quad-edge [6] data structure to represent the mesh. The algorithms, including Delaunay triangulation, Delaunay refinement, and adaptive refinement, are designed using the high-level interfaces of the Mesh class. Each algorithm is encapsulated in a separate class.

The current mesh generator in SIFFEA is sequential, which is observed the performance bottleneck of PFEM. We are also currently designing and implementing a parallel mesh generation algorithm. The basic idea is as follows. Assume that there are p processors. We first apply the quadtree [7] based technique to add some Steiner points to the input point set. Then we apply some point separator [8] or the k-d tree [9], to divide the input point set and the Steiner points into p subdomains. Each processor is then assigned a subdomain, and performs the Delaunay triangulation and refinement on it. Note that the Delaunay refinement may add Steiner points on the separator edges between subdomains. Therefore, after each processor finishes triangulation and refinement locally, it must update its point set, by merging with the Steiner points on its adjoining separator edges from other processors; then it resumes the Delaunay refinement. This process repeats until no new points are added. Notice that the Steiner points added by the quadtree technique roughly make the points set well-spaced, thus it reduces the number of Steiner points added on the separator edges later on and hence reduce communication.

3   Mesh Partition and Distribution

After mesh generation, the mesh must be distributed across processors, so that in the subsequent steps, the linear system can be assembled and solved in parallel. In this section, we describe our approaches for mesh partition and distribution.

3.1   Mesh Partition

Mesh partition is to decompose the mesh into roughly equal size of subdomains while minimizing the cut-edges between the subdomains. Its purpose is to achieve better load balance and reduce communication overhead during linear system assembly and solve. Mesh partition is typically done using a graph partitioning algorithm. In our current implementation, we use ParMetis [10], which is a fast parallel graph partitioning algorithm, to carry out the work.

The ParMetis subroutine for graph partitioning requires the input of a distributed graph. In particular, the mesh nodal graph must be distributed across processors before calling ParMetis. If a sequential mesh generator is used, we accomplish this by assigning vertices in blocks to processors. That is, if n is the number of vertices, and p is the number of processors, then vertex i is assigned to processor $i/\lceil n/p \rceil$. Note that ParMetis only requires vertex adjacency list. To collocate the mesh and utilize memory efficiently, we distribute the vertex coordinates and the element connectivities along with the vertex adjacency list. The initial element partition is determined as follows: We assign an element to the owner of its second largest vertex. This simple heuristic has the advantage of not requiring extra communication to determine the element partition, and it yields fairly even distribution of elements in practice.

After the initial distribution, the ParMetis subroutine ParMetisPartKWay() is then called, which returns the new processor assignment of the locally owned vertices on each processor. This assignment will determine the partition of the global stiffness matrix, as discussed in Sec. 4.2. Note that an element partition is also needed by the linear system assembler. Again, the element partition can be derived from the vertex partition as above.

3.2 Data Redistribution

After determining vertex and element partitions, we must ship vertices and elements to their assigned processors. We refer to this stage as *data redistribution*. In particular, four types of data must be communicated at this stage, and they all have all-to-all communication patterns.

First of all, each processor must gather the vertices assigned to it, which includes gathering the global indices, the coordinates, and the adjacency list of each vertex. Note that the adjacency list will be used by memory management of the global matrix. Since a processor is generally assigned nonconsecutive vertices, to facilitate the subsequent steps, we compute a vertex renumbering such that the vertices on each processor are consecutive in the new number system.

Secondly, we ship the element connectivities to the assigned processors. Since the element assignments were based on the old number system, the connectivities are mapped to the new number system after shipped to the owner processors.

One or more vertices of a straddling element on the cut-edges are not local to the element's owner processor. These nonlocal vertices are called the *ghost points,* which are needed by the element matrix and vector computation. In the third step, each processor determines and gathers the ghost points.

Finally the boundary vertices and their marks are shipped to the processors. A boundary vertex must be sent to a processor, if it is owned or is a ghost point of that processor. Since the number of boundary vertices is generally small, a broadcast is used instead of vector scatters, to simplify communication pattern. After the broadcasting, each processor deletes the unneeded boundary vertices.

4   Linear System Assembly

The next step of solving FEM is to assemble and solve a system of linear equations. Our current implementation of SIFFEA has adopted a matrix-based approach. Namely, we assemble a global stiffness matrix and a global load vector from the element matrices and vectors, when solving the linear system. SIFFEA features a novel design for computing the element matrices and vectors and for assembling the global matrix. This design maximizes code reuse and reduces application code development time. This section presents the rationales behind our design. For the solution of the linear system, SIFFEA currently employs the linear solves in PETSc, which includes a variety of iterative solvers and rich functionality for managing matrices and vectors. For more information about PETSc, readers are referred to [11].

4.1   Element Matrix and Vector

Solving a boundary value problem using FEM includes two distinct transformations of its mathematical formulation: the differential equation is first transformed into an integral form (a.k.a. weak form), and then the integral form is transformed into a matrix and vector form. The first transformation demands knowledge about the application, and thus is best handled by the user. The second transformation, on the other hand, is application independent, but tedious and time consuming. Therefore, an ideal interface of a system for FEM should have the flexibility for the user to specify the integral form, but require minimum user's effort to transform the integral form into matrix form.

With the above goals, we choose the integral form as the mathematical modeling input of SIFFEA, and design tools for computing the element matrices and vectors from the integral form. We categorize the tasks of computing element matrices into three levels. The lowest level is the evaluation of values and derivatives of the shape functions at the Gauss points of an element. Since there are only a small numbers of element types that are widely used, we can develop an element library for these common elements. The second level is the numerical integration of a *term* in an integral form to compute a *partial element matrix*, where a term is a product of functions. The number of possible terms is also small, and hence an integrator library can be built on top of the element library. The highest level is the summation of partial matrices into an element matrix. The same categorization applies for computing element vectors. The user typically need write applications only in the highest level using the integrator library.

We now illustrate this idea through a concrete example. Assume the input integral form is

$$\int_{\Omega} (\nabla u \cdot \nabla v + auv) \, dx \, dy = \int_{\Omega} fv \, dx \, dy,$$

where Ω denotes the domain, u and v are the base and test functions respectively, a is a constant, and f is a smooth function on Ω. We rewrite the left-hand side as a sum of two integrals, and we get

$$\int_{\Omega} \nabla u \cdot \nabla v \, dx \, dy + a \cdot \int_{\Omega} uv \, dx \, dy = \int_{\Omega} fv \, dx \, dy.$$

We can now use the integrators provided by SIFFEA to compute each integral.

There are three types of integrators. The first type integrates a term over an element and returns a partial element matrix, which are used to compute stiffness matrices. The second type also integrates a term over an element but returns a vector, used to compute load vectors. The last type integrates a term along marked boundaries of an element and returns a vector, used for applying boundary conditions, where the marks are provided by the mesh generator based on the users' input geometric specification. Each integrator type may have several functions, one for each valid term. The interface prototypes of these integrators are given as follows.

```
template <class Element> Matrix aMatIntegrator(const Element&,Fn*);
template <class Element> Vector aVecIntegrator(const Element&,Fn*);
template <class Element> Vector aBndIntegrator(const Element&,Fn*);
```

Note that these integrators are all template functions, because for a given term, the algorithms for computing its integral is the same for different elements. These integrators all have two input parameters: an object of type Element, which will be discussed shortly, and a function pointer. The function pointer points to a user function. A null pointer indicates an identity function.

Return to the above example. Let the integrators for the terms of the integral form be integrate_dudv(), integrate_uv(), and integrate_v() respectively. When translated into C++, the integrate form looks as follows.

```
template <class Element>
class ElmIntegralForm {
public:
    Matrix formELementMatrix(const Element& e){
        return integrate_dudv(e, 0)+a*integrate_uv(e, 0);
    }
    Vector formELementVector(const Element& e){
        return integrate_v(e, f);
    }
    \\ Other functions, such as Dirichlet boundary conditions,
    \\ and user function f
};
```

Readers can easily recognize an one-to-one correspondence between the terms in the integral form and in the application code. We encapsulate user's code in a parameterized class, which is consistent with the integrators. The choice of element type is then independent of mathematical specification, as it should be.

An Element class encapsulates the nodal coordinates, shape functions, and Gauss points of an element. Each type of finite element has a corresponding Element class, but all having the same interface. Note that only the linear system assembler need ever create instances of an Element class using the topological and geometrical data from the mesh data structure. In the member functions of the user class ElmIntegralForm, an Element object is passed in as a parameter. The user, however, need only pass the object to the integrators as a black box. The integrators then compute numerical integrations over the element through its public interfaces. An Element class contains public functions for computing the value and derivatives of the shape functions and the determinant of the Jacobian of the element map at the Gauss points, and for retrieving the weights at the Gauss points.

4.2   Global Matrix and Vector Assembly

Recall that in the global stiffness matrix, each row/column (or row/column block) corresponds to a vertex in the mesh, and there is a nonzero at row i

and column j if vertex i and j are adjacent in the mesh. It then follows that the global matrix is symmetric, and its nonzero pattern can be determined statically. SIFFEA employs the vertex adjacency information to preallocate memory space for the global matrix, to minimize the number of memory allocations.

We partition the global matrix onto processors by rows. In particular, a processor owns a row if it owns the corresponding vertices of that row. Linear system assembly works completely under the new vertex number system, so that each processor owns some consecutive rows of the global matrix. The global vector is also partitioned similarly.

To achieve the best scalability, the global matrix and vector are assembled on all processors concurrently. Each processor is in charge of constructing the element matrices and vectors of its locally owned elements, by calling the member function of user class ElmIntegralForm. Since all coordinates needed are already cached in the local memory, computing element matrices does not introduce extra communication. Subsequently, the element matrices and vectors are then assembled into the global matrix and vector with moderate amount of communication.

4.3   Boundary Condition Adjustment

After the element matrices are assembled into the global matrix, the global matrix is singular, and the boundary conditions must be applied. In the actual implementation, we adjust the boundary condition during matrix assembly for the best performance. For clarity, we discuss the boundary conditions separately.

The Neumann boundary conditions are handled in the integral form implicitly by the means of boundary integration. For Dirichlet conditions, we adopted a simple approach, in which each row corresponding to a Dirichlet boundary vertex is replaced by the boundary condition at that vertex. To preserve the symmetry of the global matrix, the off-diagonal nonzeros in the column corresponding to that vertex is also eliminated and moved to the right-hand side. This method has the advantages of not changing the number of equations, and hence simplifies the implementation. However, it suffers from a tiny amount of redundant computation. In terms of the user interface, a member function formDirichletBC() must be provided by the user in the class ElmIntegralForm to compute the boundary condition of a point based on its coordinates and mark.

5   Error Estimation and Adaptive Mesh Refinement

After solving the linear system in parallel, we obtain an initial numerical solution u. The initial mesh M does not guarantee a good solution, and hence it must be refined properly [12]. In this section, we consider the issues regarding to adaptive mesh refinement, and the algorithms for it.

A spacing function such as el_M and lfs_M [4] specifies how fine a mesh should be at a particular region. This function can be derived from the previous numerical results or from the local geometry feature. The spacing function for a well-shaped mesh should be smooth in the sense that it changes slowly as a function of distance, i.e., it satisfies the α-*Lipschitz* condition [13].

Our spacing function is based on Li *et al.* [13]. First, a refine-coarsening factor δ_p of each point p is obtained from *a posterior* error analysis, by comparing the error vector e of the initial solution u with a scalar global error measurement ϵ. For example, if we choose e to be the magnitude of residual, and ϵ the norm of e, then δ_p can be defined as: $(1)(e_p/\epsilon)^{-1/k}$, if $l < e_p/\epsilon < L$; $(2)1/l^{1/k}$, if $e_p/\epsilon \leq l$; $(3)1/L^{1/k}$, if $e_p/\epsilon \geq L$; where k is the accuracy order of the elements, and $0 < l \leq 1 \leq L$. The new spacing function h is then obtained from δ_p and previous spacing[13].

The mesh is then adaptively refined according to h. We would like to use the structure of the current mesh M as much and as efficiently as possible. The first step of our algorithm is to compute a maximum spacing function f that satisfies the new spacing requirement h. Then, we generate a β-sphere-packing of the domain with respect to f using the following procedure [13].

Let $S_1 = \{B(p, f(p)/2)|p \in M\}$. Then sample points in every triangle of the mesh and let S_2 be the set of these spheres defined by them. The spheres in $S_1 \cup S_2$ are ordered as the following: First the spheres centered on the boundary; then, all other spheres in S_1 in increasing order of radii; then followed by all spheres in S_2 in increasing order of radii. The intersection relation of spheres defines a *Conflict Graph* (*CG*). Let S be the set of spheres which form the *Lexical-First Maximal Independent Set* (*LFMIS*) [13] of *CG*. M' is then the Delaunay triangulation of the centers of the spheres in S. The generated mesh is well shaped and the size is within a constant factor of the optimal mesh [13].

We are currently developing a parallel adaptive mesh refinement algorithm, in which the same idea as the parallel mesh generation is to be used.

6   Experimental Study

The input geometry of our testing case is shown in Fig. 2. The testing mesh is finer than the one shown in Fig. 2, containing 193881 vertices, 535918 edges, and 342033 triangles. The minimum minimum angle of the triangles is 20.1°. The model being solved is a Laplace equation. The boundary conditions are set as follows. The innermost circle has no displacement, and the outermost circle has unit displacement in their normal directions. All the other boundaries have zero natural boundary condition. The computational solution of this sample problem is visualized in Fig. 3, which plots the displacement vectors of the mesh vertices.

We conducted some preliminary experiments on an Origin 2000 with 128 processors at NCSA, which is a distributed shared memory machine. Each processor is a 250 MHz R10000. The native implementation of MPI on Origin was used during the test. The sparse linear solver used was Conjugate Gradient method, where the absolute and relative error tolerances were both set at 10^{-8}. The preprocessing step is done in sequential, including Delaunay triangulation and refinement.

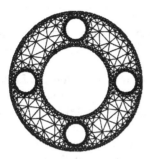

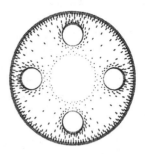

Fig. 2. The input geometry of the testing problem.

Fig. 3. Displacement vectors.

We present the execution times of data distribution, mesh partitioning, matrix assembly, and linear solver in Table 1. The results show that the execution times of data distributions, which involve intensive communication, do not change much with the number of processors. The mesh partitioner slows down after 32 processors, maybe due to the following two reasons: (1) The initial vertex distribution is essentially random, which causes large amount of communication on large number of processors; (2) the difficulty of the partitioning problem itself increases with the number of the processors. For the similar reasons, the data redistribution also has no speedups for more than 16 processors. The linear system assembler and the linear solver achieve steady speedups up to 128 processors. To isolate the slowdowns of ParMetis and data redistribution, we report the speedups of both with and without ParMetis and data redistribution. The results clearly demonstrate the necessity of parallel mesh generation: After parallel mesh generation, each processor will contain a mesh of a subdomain, which already gives a fairly good partition; then a cheaper partition refinement and data redistribution can be used.

Table 1. Performance results on Origin 2000 at NCSA.

Time in seconds	1	2	4	8	16	32	64	128
Initial distribution	-	0.41	0.35	0.51	0.35	0.48	0.61	0.46
ParMetis (M)	-	2.0	0.94	0.68	0.84	2.7	23.5	59.0
Redistribution (R)	-	0.50	0.27	0.26	0.25	0.31	1.4	12.3
Linear assembler	12.9	10.7	7.2	5.2	4.5	3.9	2.2	1.4
Linear solver	269.6	186.2	83.8	47.0	17.1	16.6	6.8	5.5
Total	382.5	199.8	92.6	53.7	23.0	24.0	34.5	78.7
Speedup(w/ M,R)	-	1.9	4.1	7.1	16.6	15.9	11.1	4.9
Speedup(w/o M,R)	-	1.9	4.2	7.25	17.5	18.21	39.8	50.6

7   Related Work

A lot of work has been done on PFEM in the past few years. Various approaches have been used by researchers. In this section, we summarize some features of related work and highlight their differences from SIFFEA.

Archimedes [14], developed at the CMU, is an integrated tool set for solving PDE problems on parallel computers. It employs algorithms for sequential mesh generation, partitioning, and unstructured parallel computation as a base for numerical solution of FEM. The central tool is Author, a data parallelizing compiler for unstructured finite element problems. The user writes a complete sequential FEM solver, including linear system assembly and solve, and then Author automatically parallelizes the solver. Archimedes is written in C.

The Prometheus library [15], is a parallel multigrid-based linear system solver for unstructured finite element problems developed at Berkeley. It takes a mesh distributed across multiple processors, and automatically generates all the coarse representations and operators for standard multigrid methods. Prometheus does not integrate mesh generation and adaptive mesh refinement.

There is also some work in the area of parallel adaptive mesh refinement. PYRAMID[16], developed at NASA and written in Fortran 90, uses the edge bisection to refine a mesh in parallel. To achieve load balancing, it estimates the size of the refined mesh and redistribute the mesh based on a weighted scheme before the actual refinement is performed. SUMMA3D[17] developed at ANL uses the *longest edge bisection method* to refine a mesh. It applies its own mesh partition algorithm after mesh refinement.

8   Conclusion and Future Work

We have described SIFFEA, an integrated framework for PFEM, which synthesizes several widely used components in FEM. SIFFEA features a novel design for specifying the mathematical model which maximizes code reuse and simplifies the user interface. As an implicit parallel environment, SIFFEA hides all internal communications from the user. The preliminary experimental results have achieved good overall speedups up to 32 processors, and reasonable speedups up to 128 processors for our data distribution and matrix assembly codes. We are underway of designing and implementing algorithms for parallel mesh generation and refinement, to achieve better overall scalability. We also plan to extend the framework to time-dependent and nonlinear problems, and to provide interfaces with multigrid and direct linear solvers.

Acknowledgments

This work started as a course project with Professor Laxmikant Kale, who has encouraged us to work on the problem continuously and to put it into publication. We thank Jiantao Zheng from the TAM Department of UIUC for helpful discussions on theories of finite element methods. We thank the anonymous referees for their valuable comments. The authors are supported by the Center for Simulation of Advanced Rockets funded by the U.S. Department of Energy under Subcontract B341494.

References

1. I. Babuška and A. K. Aziz. On the angle condition in the finite element method. *SIAM J. Numer. Anal.*, 13(2):214–226, 1976.
2. G. Strang and G. J. Fix. *An Analysis of the Finite Element Method.* Prentice-Hall, 1973.
3. R. Sibson. Locally equiangular triangulations. *Computer Journal*, 21:243–245, 1978.
4. J. Ruppert. A Delaunay refinement algorithm for quality 2-dimensional mesh generation. *Journal of Algorithms*, 18(3):548–585, 1995.
5. X. Y. Li, S. H. Teng, and A. Üngör. Biting: advancing front meets sphere packing. *the International Journal of Numerical Methods in Engineering*, 1999. to appear.
6. L. J. Guibas and J. Stolfi. Primitives for the manipulation of general subdivisions and the computation of Voronoi diagrams. *ACM Trans. Graphics*, 4(2):74–123, 1985.
7. Marshall W. Bern, David Eppstein, and Shang-Hua Teng. Parallel construction of quadtrees and quality triangulations. In Frank K. H. A. Dehne, Jörg-Rüdiger Sack, Nicola Santoro, and Sue Whitesides, editors, *Algorithms and Data Structures, Third Workshop*, volume 709 of *Lecture Notes in Computer Science*, pages 188–199, Montréal, Canada, 1993. Springer.
8. G. L. Miller, S. H. Teng, W. Thurston, and S. A. Vavasis. Geometric separators for finite element meshes. *SIAM J. Scientific Computing*, 19(2):364–384, 1998.
9. J.L. Bentley. Multidimensional binary search trees used for associative searching. *Communication of the ACM*, 18(9), 1975.
10. G. Karypis and V. Kumar. Parallel multilevel k-way partitioning scheme for irregular graphs. TR 96-036, Computer Science Department, University of Minnesota, Minneapolis, MN 55454, 1996.
11. S. Balay, W. D. Gropp, L. C. McInnes, and B. F. Smith. PETSc 2.0 users manual. Technical Report ANL-95/11, Argonne National Laboratory, 1998.
12. I. Babuška, O. C. Zienkiewicz, J. Gago, and E. R. de A. Oliveira, editors. *Accuracy Estimates and Adaptive Refinement in Finite Element Computations.* Wiley Series in Numerical Methods in Engineering. John Wiley, 1986.
13. X. Y. Li, S. H. Teng, and A. Üngör. Simultaneous refinement and coarsening: adaptive meshing with moving boundaries. *Journal of Engineering with Computers*, 1999. to appear.
14. J. R. Shewchuk and D. R. O'Hallaron. Archimedes home page. http://www.cs.cmu.edu/~quake/arch.html, 1999.
15. M. Adams, J. W. Demmel, and R. L. Taylor. Prometheus home page. http://www.cs.berkeley.edu/~madams/Prometheus-1.0, 1999.
16. J. Lou, C. Norton, V. Decyk, and T. Cwik. Pyramid home page. http://www-hpc.jpl.nasa.gov/APPS/AMR/, 1999.
17. Lori Freitag, Carl Ollivier-Gooch, Mark Jones, and Paul Plassmann. SUMMA3D home page. http://www-unix.mcs.anl.gov/~freitag/SC94demo/, 1999.

Overture: Object-Oriented Tools for Applications with Complex Geometry

David L. Brown, William D. Henshaw, and Dan Quinlan

Lawrence Livermore National Laboratory, Livermore, CA, USA

Abstract. The *Overture* framework is an object-oriented environment for solving partial differential equations in two and three space dimensions. It is a collection of C++ libraries that enables the use of finite difference and finite volume methods at a level that hides the details of the associated data structures. *Overture* can be used to solve problems in complicated, moving geometries using the method of overlapping grids. It has support for grid generation, difference operators, boundary conditions, data-base access and graphics. In this paper we briefly present *Overture* , present some of the newer grid generation capabilities, and discuss our approach toward performance within *Overture* and the A++P++ array class abstractions upon which *Overture* depends, this work represents some of the newest work in *Overture* . The results we present show that the abstractions represented within *Overture* and the A++P++ array class library can be used to obtain application codes with performance equivalent to that of optimized C and Fortran 77. Further, the preprocessor mechanism for which approach we present results, is general in its application to any object-oriented framework or application and is not specific to *Overture* .

1 Introduction

The *Overture* framework is a collection of C++ libraries that provide tools for solving partial differential equations. *Overture* can be used to solve problems in complicated, moving geometries using the method of overlapping grids (also known as overset or Chimera grids). *Overture* includes support for geometry, grid generation, difference operators, boundary conditions, data-base access and graphics.

An overlapping grid consists of a set of logically rectangular grids that cover a domain and overlap where they meet. This method has been used successfully over the last decade and a half, primarily to solve problems involving fluid flow in complex, often dynamically moving, geometries. Solution values at the overlap are determined by interpolation. The overlapping grid approach is particularly efficient for rapidly generating high-quality grids for moving geometries. As the component grids move only the boundary points to be interpolated change, the grid points do not have to be regenerated. The component grids are structured so that efficient and fast finite-difference algorithms can be utilized. We use adaptive mesh refinement to resolve fine features of the solution. [1,2]. The design

S. Matsuoka et al. (Eds.): ISCOPE'99, LNCS 1732, pp. 96–107, 1999.

of *Overture* has evolved over the past 15 years or so from the Fortran 77 based CMPGRD [3] environment to the current C++ version [4,5]. Although the Fortran 77 implementation was used for complicated three-dimensional adaptive and moving grid computations, the programs were difficult to write and maintain. *Overture* was designed to have at least all the functionality of the Fortran code but to be as easy as possible to use; indeed, an entire PDE solver on an overlapping grid can be written on a single page (see Sect. 3).

Overture is an object-oriented framework. In the past a typical Fortran code would use a procedural model where subroutines and functions are the fundamental building blocks and data is passed to and from these procedures. In *Overture* the fundamental building blocks are objects such as grids and grid functions. These objects can be manipulated at a high level. Details of the implementation, such as how a grid is stored, are hidden from the user, but importantly all data is accessible to permit the optional use of the data directly as required for lower level optimizations. Within *Overture* , the A++P++ array class library is used both internally and within the user interface, thus the performance of *Overture* greatly depends upon the performance of operations expressed within the A++/P++ array class (array objects). In this paper, we present a preprocessor based approach toward performance within *Overture* along with some simple results. The point of the preprocessor is to provide a mechanism to leverage the semantics of the object-oriented abstractions within A++P++ and use them to drive the automated introduction of highly optimized transformations.

There are a number of other very interesting projects developing scientific object-oriented frameworks. These include the SAMRAI framework for structured adaptive mesh refinement[6], PETSc (the Portable Extensible Toolkit for Scientific Computation)[7], POOMA (Parallel Object Oriented Methods and Applications)[8], Blitz++[9] and Diffpack[10].

2   The Overture Framework

The main class categories that make up *Overture* are as follows:

- **Arrays** [11]: describe multidimensional arrays using A++/P++. A++ provides the serial array objects, and P++ provides the distribution and interpretation of communication required for their data parallel execution.
- **Mappings** [12]: define transformations such as curves, surfaces, areas, and volumes. These are used to represent the geometry of the computational domain.
- **Grids** [13,14]: define a discrete representation of a mapping or mappings. These include single grids, and collections of grids; in particular composite overlapping grids.
- **Grid functions** [14]: storage of solution values, such as density, velocity, pressure, defined at each point on the grid(s). Grid functions are derived from A++/P++ array objects.
- **Operators** [15,16]: provide discrete representations of differential operators and boundary conditions

- **Grid generation** [17]: the Ogen overlapping grid generator automatically constructs an overlapping grid given the component grids.
- **Plotting** [18]: a high-level interface based on OpenGL allows for plotting *Overture* objects.
- **Adaptive mesh refinement**: The AMR++ library is an object-oriented library providing patch based adaptive mesh refinement capabilities within *Overture* .

Solvers for partial differential equations, such as the *OverBlown* solver are available from the *Overture* Web Site.

2.1 Array Operations

A++ and P++ [11,19] are array class libraries for performing array operations in C++ in serial and parallel environments, respectively.

A++ is a *serial* array class library similar to FORTRAN 90 in syntax, but not requiring any modification to the C++ compiler or language. A++ provides an object-oriented array abstraction specifically well suited to large scale numerical computation. It provides efficient use of multidimensional array objects which serves to both simplify the development of numerical software and provide a basis for the development of parallel array abstractions. P++ is the *parallel* array class library and shares an identical interface to A++, effectively allowing A++ serial applications to be recompiled using P++ and thus run in parallel. This provides a simple and elegant mechanism that allows serial code to be reused in the parallel environment.

P++ provides a data parallel implementation of the array syntax represented by the A++ array class library. To this extent it shares a lot of commonality with FORTRAN 90 array syntax and the HPF programming model. However, in contrast to HPF, P++ provides a more general mechanism for the distribution of arrays and greater control as required for the multiple grid applications represented by both the overlapping grid model and the adaptive mesh refinement (AMR) model. Additionally, current work is addressing the addition of task parallelism as required for parallel adaptive mesh refinement.

Here is a simple example code segment that solves Poisson's equation in either a serial or parallel environment using the A++/P++ classes. Notice how the Jacobi iteration for the entire array can be written in one statement.

```
// Solve u_xx + u_yy = f by a Jacobi Iteration
Range R(0,n)                    // define a range of indices: 0,1,2,...,n
floatArray u(R,R), f(R,R)       // declare two two-dimensional arrays
f = 1.; u = 0.; h = 1./n;       // initialize arrays and parameters
Range I(1,n-1), J(1,n-1);       // define ranges for the interior

for( int iteration=0; iteration<100; iteration++ )
  //data parallel
  u(I,J) = .25*(u(I+1,J)+u(I-1,J)+u(I,J+1)+u(I,J-1)-f(I,J)*(h*h));
```

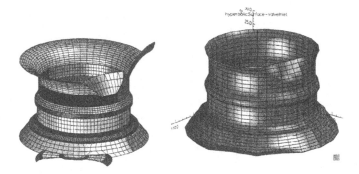

Fig. 1. Hyperbolic surface grid generation is used to generate a smooth surface grid over a surface coming from a CAD package.

2.2 Grid Generation

Overture has support for the creation of overlapping grids for complex geometries. The process of generating an overlapping grid consists of two basic steps. First, several component grids are generated. Each component grid represents a portion of the geometry. The component grids must overlap but otherwise can be created locally. *Overture* provides a collection of Mapping classes that can be used to generate component grids including splines, NURBS, bodies of revolution, hyperbolic grid generation, elliptic grid generation, trans-finite interpolation and so on. In addition we are working on methods for reading files generated by CAD programs and generating grids. Figure (1) shows how hyperbolic grid surface grid generation can be used to generate a single smooth grid over a CAD surface described by a collection of trimmed NURBS. This is accomplished with the aid of the SURGRD hyperbolic surface grid generator[20].

Second, the overlapping grid is constructed using the Ogen grid generator. This latter step consists of determining how the different component grids interpolate from each other, and in removing grid points from holes in the domain, and removing unnecessary grid points in regions of excess overlap. Ogen requires a minimal amount of user input. The grids in Fig. (2) were all created with Ogen and represent some of the new grid generation capabilities within *Overture* .

3   Writing PDE Solvers

This example demonstrates the power of the *Overture* framework by showing a basically complete code that solves the partial differential equation (PDE)

$$u_t + au_x + bu_y = \nu(u_{xx} + u_{yy})$$

on an overlapping grid.

The PlotStuff object is used to interactively plot contours of the solution at each time step[18].

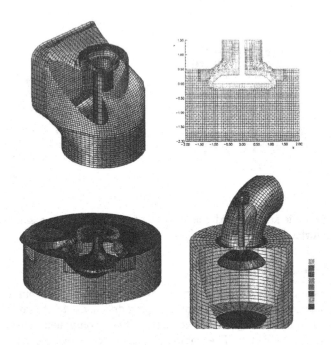

Fig. 2. Sample 2D and 3D overlapping grids generated with the Ogen grid generator.

```
int main() \{
  CompositeGrid cg;                          // create a composite grid
  getFromADataBaseFile(cg,"myGrid.hdf");     // read the grid in
  floatCompositeGridFunction u(cg);          // create a grid function
  u=1.;                                      // assign initial conditions
  CompositeGridOperators op(cg);             // create operators
  u.setOperators(cg);
  PlotStuff ps;                              // make an object for plotting
  // --- solve a PDE ----
  float t=0, dt=.005, a=1., b=1., nu=.1;
  for( int step=0; step<100; step++ )
  \{
    u+=dt*( -a*u.x()-b*u.y()+nu*(u.xx()+u.yy()) );
    t+=dt;
    u.interpolate();                         // interpolate overlapping boundaries
    // apply the BC u=0 on all boundaries
    u.applyBoundaryCondition(0,dirichlet,allBoundaries,0.);
    u.finishBoundaryConditions();
    ps.contour(u);                           // plot contours of the solution
  \}
  return 0;
\}
```

The example solves the time-dependent equation explicitly. Other class libraries within the *Overture* framework simplify the solution of elliptic and parabolic equations, the linear systems generated can be solved using any of numerous numerical methods as appropriate including multigrid, and methods made available within a number of external dense and sparse linear algebra packages including PETSc, and others. These are wrapped into the elliptic solver library (Oges) within *Overture* .

4 Approach to Performance

The execution of array statements involves inefficiencies stemming from several sources and the problem has been well documented, by many researchers[21,9,22]. Our approach to performance within *Overture* is to use a preprocessor to introduce optimizing source-to-source transformations. The C++ source-to-source preprocessor is built using ROSE; a tool we have designed and implemented to build application specific preprocessors.

ROSE is a programmable source-to-source transformation tool built on top of SAGE[23] for the optimization of C++ object-oriented frameworks. While we have specific goals for this work within *Overture* , ROSE applies equally well to any other object-oriented framework.

A common problem within object-oriented C++ scientific computing is that the high level semantics of abstractions introduced (e.g. parallel array objects) are ignored by the C++ compiler. Classes and overloaded operators are seen as unoptimizable structures and function calls. Such abstractions can provide for particularly simple development of large scale parallel scientific software, but the lack of optimization greatly affects performance and utility. Because C++ lacks a mechanism to interact with the compiler, elaborate mechanisms are often implemented within such parallel frameworks to introduce complex template-based and/or runtime optimizations (such as runtime dependence analysis, deferred evaluation, runtime code generation, etc.). These approaches are however not satisfactory since they are often marginally effective, require long compile times, and/or are not sufficiently robust.

Preprocessors built using ROSE have a few features that stand out:

1. A hierarchy of grammars are specified as input to ROSE to build (tailor) the preprocessor specific to a given object-oriented application, library, or framework. ROSETTA, a code generator we have designed and implemented, is used to generate an implementation of the grammars that are used internally. The hierarchy of grammars (and their implementations) are used to construct separate program trees internally, one program tree per grammar, each representing the user's application. The program trees are edited as required to replace selected subtrees with other subtrees representing a specific transformation. Quite complex criteria may be used to identify where transformations may be applied, this mechanism is superior to pattern-recognition of static subtrees within the program tree because it is more general and readily tailored.

2. Transformations are specified which are then built into the user application automatically where appropriate. The mechanism is designed to permit the automated introduction of particularly complex transformations (such as the cache based transformations specified in [24], space does not permit an elaboration of this.

3. To simplify the debugging, preprocessor's output (C++ code) is formatted identical to the input application code (except for transformations that are introduced, which have a default formatting). Numerous options are included to tailor the formatting of the output code and to simplify working with either its view directly within the debugger or its reference to the original application source within the debugger. Comments and all C preprocessor (cpp) control structures are preserved within the output C++ code.

4. The design of ROSE is simplified by leveraging both SAGE 2 and the EDG[25] C++ front-end. EDG supplies numerous vendors with the C++ front-end for their compiler and represents the current best implementation of C++. In principle this permits the preprocessors built by ROSE to address the complete C++ language (as implemented by the best available front-end). Modifications have been made to SAGE 2 to permit portability and allow us to fulfill on a complete representation of the language. By design, we leverage many low level optimizations provided within modern compilers while focusing on higher level optimizations largely out of reach because traditional approaches can not leverage the semantics of high level abstractions. In doing so, we slightly blur the distinction between a library or framework, a language, and a compiler. But, because we leverage several good quality tools the implementation is greatly simplified.

The approach is different from other open C++ compiler approaches because it provides a mechanism for defining high level grammars specific to an object-oriented framework and a relatively simple approach to the specification of large and complex transformations. A requirement for representing the program tree within different user defined grammars is to have access to the full program tree, this is not possible (as we best understand) within the OpenC++[26] research work. By using SAGE 2 and ROSE the entire program tree, represented in each grammar, is made available; this permits more sophisticated program analysis (when combined with the greater semantic knowledge of object-oriented abstractions) and more complex transformations. We believe that the techniques we have developed greatly complement the approaches represented within OpenC++, in particular the Meta object mechanism represented within that work. That SAGE is in many ways similar to the MPC++[27] work, we believe we could have alternatively built off of that tool in place of SAGE (though this is not clear). However, since SAGE 2 uses the EDG front-end we expect this will simplify access to the complete C++ language. MPC++ addresses more of the issues associated with easily introducing some transformations than SAGE, but not of the complexity that we require for cache based transformations[24]. Each represent only a single grammar (the C++ grammar) and this is far too complex (we believe) a starting point for the identification of where sophisticated transformations can be introduced. The overall compile-time optimization goals are related to ideas put forward by Ian Angus[28], but with numerous distinguishing points:

1. We have decoupled the optimization from the back-end compiler to simplify the design.
2. We have developed hierarchies of grammars to permit arbitrarily high level abstractions to be represented with the greatest simplicity with the program tree. The use of multiple program trees (one for each grammar) serves to organize high level transformations.
3. We provide a simple mechanism to implement transformations.
4. We leverage the semantics of the abstractions to drive optimizations.
5. We have implemented and demonstrated the preprocessor approach on several large numerical applications.

Finally, because ROSE is based ultimately (through SAGE) upon the EDG C++ front-end, the full language is made available; consistent with the best of the commercial vendor C++ compilers which most often use the same EDG C++ front-end internally. However, some aspects of the complete support of C++ within SAGE are incomplete.

5   Results

Within our results we consider the following trivial five point stencil:

```
A(I,J)  =   c*( B(I-1,J)+B(I+1,J)+B(I,J)+B(I,J+1)+B(I,J+1));
```

In this code fragment, A and B are multidimensional array objects (distributed across multiple processors if P++ is used). In this example, I and J are **Range** objects that together specify an index space of the arrays A and B.

Figure 3 shows the range of performance associated with different size arrays for the simple five point stencil operator on the Sun Ultra and Dec Alpha machines. The Sun Ultra was selected because it is a commonly available computer, the Dec Alpha was selected because its cache design is particularly aggressive and as a result it exemplifies the hardest machine to get good cache performance. The results are in no way specific to this statement, moderate size applications have been processed using preprocessors built with ROSE. The results compare the ratios of A++ performance with and without the use of the ROSE preprocessor to that of optimized C code. The optimized C code takes full advantage of the bases of the arrays being identical and the unit strides, the A++ implementation does not, these very general subscript computations within the array class implementation are compared to very specific and highly optimized subscript computations within the C code. This exaggerates the poorer performance of the A++ statements, we do this to make clear that the performance of the code output from the ROSE preprocessor is in fact highly optimized and is made specific to the common bases of the operands (determined at compile time) and

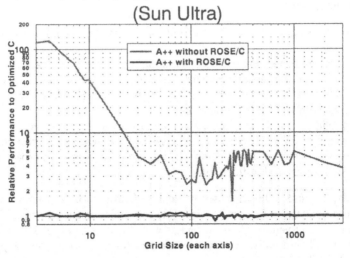

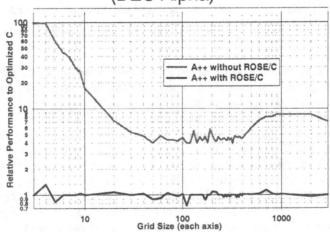

Fig. 3. The use of a preprocessor (built using ROSE) can overcome the performance degradation associated with binary evaluation of array operands. These results show the use of ROSE with A++ and how the performance matches that of optimized C code using the `restrict` keyword (ratio = 1). It has been shown previously that this is equal to Fortran 77 performance. More sophisticated cache-based transformations are also possible.

the unit stride (determined at runtime). Our results show the relative difference that it makes to compare such results. The resulting performance using ROSE is nearly identical to that of the optimized C code (ratio = 1), this is not surprising since the preprocessor transformation replaces the array statement with the equivalent C code (highly optimized, and using restrict pointers where they are supported).

A++ supports expression templates but this data is not presented here, in general the expression template will approach the C performance within 90% for short expressions and sufficiently large arrays. The combination of expression templates with deferred evaluation reduces this to about 70% as reported in [22] likely because of the required extra level of indirection to the data required by the deferred evaluation mechanism (it is not clear if this will be fixed)[1].

An important distinguishing point between the two approaches is that within larger applications the compile times are several orders of magnitude less for the preprocessor approach since expression templates are not used[21]. In practice the time taken to pre-process an application is even much less than the compile time where no templates are used (expression templates or otherwise) (a few seconds, and is not noticeable). This is not surprising since the preprocessing consists of only a few of the steps taken internally within a compiler, and excludes the most time consuming back-end optimization (to build the object code).

6   Conclusions

Overture is capable of addressing the complexity of numerous difficult sorts of simulations within scientific computing. We have demonstrated that powerful abstractions can be developed that can greatly simplify the development of previously complex (intractable) simulations. New features within the grid generator in *Overture* have made much more complicated grids possible. While the abstractions presenting within *Overture* are the principle motivation for its use, the performance of *Overture* is critical and is dominated by the performance of the A++P++ array class. Many years of work have gone into the development of optimization techniques for the array class library. The preprocessor approach is by far the most successful so far, however more work remains to make preprocessors easier to build and more robust.

The results we have presented demonstrate the optimization of array class statements. All sizes of arrays benefit, their processing with ROSE makes each equivalent to the performance of optimized C code (using restrict). Previously in [29] we showed that this is equivalent to FORTRAN 77 performance.

Expression Templates is an alternative mechanism that can be used to optimize array statements, but the mechanism is problematic[21]. More research is required (and being done by others) to address problems within the expression template mechanism. More work is similarly required to provide improved compile-time optimization solutions.

---

[1] This was the experience with expression templates when it was combined with the deferred evaluation mechanism in A++P++.

7 Software Availability

The *Overture* framework and documentation is available for public distribution from the web site, http://www.llnl.gov/casc/Overture. The **OverBlown** flow solver is also available.

References

1. M. J. Berger and P. Colella, *Local adaptive mesh refinement for shock hydrodynamics*, J. Comp. Phys., 82 (1989), pp. 64–84.
2. D. Quinlan, *Adaptive Mesh Refinement for Distributed Parallel Processors*, PhD thesis, University of Colorado, Denver, June 1993.
3. G. Chesshire and W. D. Henshaw, *Composite overlapping meshes for the solution of partial differential equations*, J. Comp. Phys., 90 (1990), pp. 1–64.
4. D. L. Brown, Geoffrey S. Chesshire, William D. Henshaw and Daniel J. Quinlan, *Overture : An Object Oriented Software System for Solving Partial Differential Equations in Serial and Parallel Environments*, Proceedings of the Eight SIAM Conference on Parallel Processing for Scientific Computing, 1997.
5. D. L. Brown, William D. Henshaw and Daniel J. Quinlan, *Overture : An Object Oriented Framework for Solving Partial Differential Equations*, Scientific Computing in Object-Oriented Parallel Environments, Springer Lecture Notes in Computer Science, 1343, 1997.
6. Xabier Garaizar, Richard Hornung and Scott Kohn, Structured Adaptive Mesh Refinement Applications Infracture, http://www.llnl.gov/casc/SAMRAI.
7. Satish Balay, William Gropp, Lois Curfman McInnes and Barry Smith, *The Portable Extensible Toolkit for Scientific Computation.* http://www.mcs.anl.gov/petsc/petsc.html.
8. Steve Karmesin et.al, *Parallel Object Oriented Methods and Applications.* http://www.acl.lanl.gov/PoomaFramework.
9. Todd Veldhuizen *Arrays in Blitz++* In *Proceeding of the Second International Symposium, ISCOPE 98*, Santa Fe, NM December 1998
10. Diffpack homepage, http://www.nobjects.com/diffpack.
11. D. Quinlan, *A++/P++ manual*, LANL Unclassified Report 95-3273, Los Alamos National Laboratory, 1995.
12. William D. Henshaw, *Mappings for Overture : A description of the mapping class and documentation for many useful mappings*, LANL unclassified report 96-3469, Los Alamos National Laboratory, 1996.
13. G. S. Chesshire, *Overture : the grid classes*, LANL unclassified report 96-3708, Los Alamos National Laboratory, 1996.
14. William D. Henshaw, *Grid, GridFunction and Interpolant classes for Overture , AMR++ and CMPGRD, user guide, version 1.00*, LANL unclassified report 96-3464, Los Alamos National Laboratory, 1996.
15. M. J. Berger and P. Colella, *Classes for finite volume operators and projection operators*, LANL unclassified report 96-3470, Los Alamos National Laboratory, 1996.
16. William D. Henshaw, *Finite difference operators and boundary conditions for Overture, user guide, version 1.00*, LANL unclassified report 96-3467, Los Alamos National Laboratory, 1996.

17. William D. Henshaw, *Ogen: an overlapping grid generator for Overture*, LANL unclassified report 96-3466, Los Alamos National Laboratory, 1996.
18. William D. Henshaw, *PlotStuff: a class for plotting stuff from Overture* , LANL unclassified report 96-3893, Los Alamos National Laboratory, 1996.
19. Lemke, M., Quinlan, D., *P++, a C++ Virtual Shared Grids Based Programming Environment for Architecture-Independent Development of Structured Grid Applications* In *Proceedings of the CONPAR/VAPP V*, September 1992, Lyon, France; published in *Lecture Notes in Computer Science*, Springer Verlag, September 1992.
20. W.M. Chan and P.G. Buning, *A Hyperbolic Surface Grid Generation Scheme and Its Applications*, AIAA paper 94-2208, 1994.
21. Bassetti, F., Davis, K., Quinlan, D. *A Comparison of Performance-enhancing Strategies for Parallel Numerical Object-Oriented Frameworks* In *Proceedings of the first International Scientific Computing in Object-Oriented Parallel Environments (ISCOPE) Conference*, Marina del Rey, California, Dec, 1997
22. Karmesin, et al. *Array Design and Expression Evaluation in POOMA II.* In *Proceeding of the Second International Symposium, ISCOPE 98*, Santa Fe, NM December 1998
23. B. Francois et. al. *Sage++: An object-oriented toolkit and class library for building fortran and c++ restructuring tools.* In *Proceedings of the Second Annual Object-Oriented Numerics Conference*, 1994.
24. Bassetti, F., Davis, K., Quinlan, D. *Optimizing Transformations of Stencil Operations for Parallel Object-Oriented Scientific Frameworks on Cache-Based Architectures* In *Proceedings of the ISCOPE'98 Conference*, Santa Fe, New Mexico, Dec 13-16 1998
25. Edison Design Group http://www.edg.com
26. Shigeru Chiba   *Macro Processing in Object-Oriented Languages*   In Proc. of Technology of Object-Oriented Languages and Systems (TOOLS Pacific '98), Australia, November, IEEE Press, 1998. more info available at: http://www.hlla.is.tsukuba.ac.jp/ chiba/openc++.html
27. Ishkawa et. al. *Design and Implementation of Metalevel Architecture in C++ - MPC++ Approach -.* In *Proceeding of Reflection'96 Conference*, April 1996 more info available at: http://pdswww.rwcp.or.jp/mpc++/mpc++.html
28. Ian Angus *Applications Demand Class-Specific Optimizations: The C++ Compiler Can Do More.* In *Proceedings of the Object-Oriented Numerics Conference*, (OONSKI) 1993
29. Bassetti, F., Davis, K., Quinlan, D. *Toward FORTRAN 77 Performance From Object-Oriented C++ Scientific Frameworks* In *Proceedings of the HPC'98 Conference*, Boston, Mass. April 5-9, 1998

Generic Programming for Parallel Mesh Problems

Jens Gerlach[1] and Mitsuhisa Sato[2]

[1] GMD-FIRST, 12489 Berlin, Germany
[2] Real World Computing Partnership, Japan

Abstract. We use the paradigm of *generic programming* together with the Standard Template Library (STL) to develop concepts and template classes that can represent finite sets and their relations as they occur in (parallel) finite element and other numerical methods. A key idea of our approach is to consider these sets as static search structures with clearly separated phases for insertions and retrievals. The resulting C++ template library *Janus* has generic **domain** and **relation** classes that allows a space-optimal and quickly traversable representation of finite element meshes and sparse matrix graphs. We give performance numbers for the parallel implementation of the Poisson equation for linear triangular elements on the Linux PC cluster of the Real World Computing Partnership. We also compare our approach with related work such as the HPC++ and POOMA projects.

1 Introduction

Our C++ template library Janus provides generic classes and an extensible conceptual framework for domains and their relations. The work at Janus has been motivated by the need of flexible and efficient representations of finite sets (e.g. triangulations and node sets) and (nearest neighbor or hierarchic) relations of finite element meshes. Here we mention some fundamental characteristics of these structures.

1. They are more complex than rectangular grids.
2. The sets and relations might evolve over the process of computation.
3. It is possible to identify clearly separated *modification* and *access* phases.

In addition we must take into account parallel platforms which often require an explicitly distributed representation of these structures.

The points 1 and 2 indicate that in this application field multidimensional arrays as they occur in the Fortran and C worlds are of little value. Moreover there are several approaches that resort to classical dynamic data structures such as lists, or trees to represent finite element meshes [1,2].

The question arises whether it is really necessary to use these highly flexible yet relatively heavy-weight and slow data structures.

S. Matsuoka et al. (Eds.): ISCOPE'99, LNCS 1732, pp. 108–119, 1999.

The key to understand that this question can be answered by "no" is the observation made in point 3. The spatial structures of many dynamic numerical problems fit into the scheme of alternating modification and access phases. We use the term Two-Phase-Property to denote this fundamental characteristic [3]. In the literature we found the concept of a *static search structure* [4] that describes the Two-Phase-Property more formally. The fundamental idea of our paper is to apply the paradigm of *generic programming* [5,6] to develop a hierarchy of static search structures that is appropriate for numeric applications. This is done in Sect. 2.

The generic domain and relation classes that are provided by Janus for sequential and parallel environments are presented in Sect. 3. The design and implementation of Janus is closely related to the Standard Template Library (STL) [7] which is probably the best known example of applying the ideas of generic programming. Since our fundamental concept is that of static search structure we can fall back on implementation strategies that proved to be useful in the area of high performance computing. As an example we mention that our irregular relation classes use the Compressed Row Storage (CRS) format.

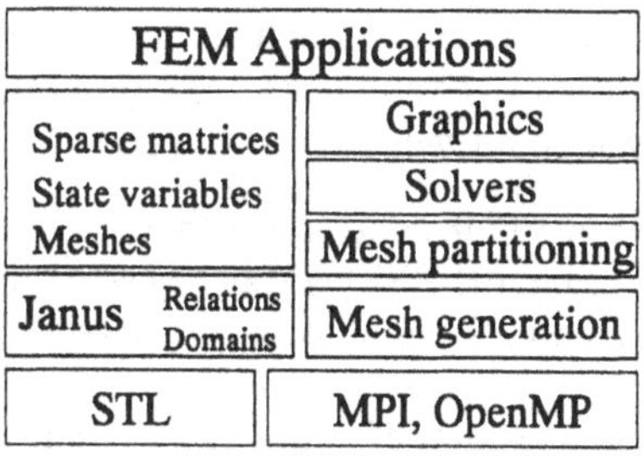

FEM Applications	
Sparse matrices	Graphics
State variables	Solvers
Meshes	Mesh partitioning
Janus Relations Domains	Mesh generation
STL	MPI, OpenMP

Fig. 1. Janus provides support for *domains* and *relations* of mesh problems.

It turns out that Janus is as a thin layer on top of the STL and communication primitives such as MPI. The components of Janus are basic building blocks for the potentially distributed sets and relations of concrete mesh methods rather than complete finite element modules. The reason for this is that concrete FEM are too diverse to provide a universal mesh structure (even when fixing the space dimension). We decided to look at what is common to basically all of them. We found that it is the need to represent a multitude of mostly irregular sets of different objects and their complex relations in sequential and distributed contexts. To write a complete FEM program more components including mesh generators, graph partitioners and linear algebra libraries shown in Fig. 1 are necessary.

How Janus can be used for the implementation of a two-dimensional FEM problem is demonstrated in Sect. 4. Before drawing conclusions and explaining our future work we compare in Sect. 5 our approach with other, well known C++ template frameworks for high performance computing, namely the Parallel STL

(PSTL) in the HPC++ project [8], the POOMA framework [9], and the Matrix Template Library (MTL) [10]. The fundamental difference is that Janus focuses on the spatial structures of numeric methods rather than on providing merely array and matrix abstractions, i.e, the algebraic objects that are defined on the spatial structures.

Readers of this papers should be familiar with the C++ programming language [11] in particular with the STL [7]. Nevertheless we think that this work is of interest to any application programmer from the field of high performance parallel computing.

2   Concepts for Domains and their Relations

For a definition of the term "generic programming" we follow D. Musser [6].

> My working definition of generic programming is "programming with concepts", where a concept is defined as a family of abstractions that are all related by a common set of requirements. A large part of the activity of generic programming, particularly in the design of generic software components, consists of concept development — identifying sets of requirements that are general enough to be met by a large family of abstractions but still restrictive enough that programs can be written that work efficiently with all members of the family.

A concept is said to be a *refinement* of another concept if it imposes more requirements. When describing the Janus framework we follow the terminology of the STL by saying that a type that satisfies the requirements of a concept C is a *model of C*.

The design and implementation of Janus is based on the STL. It reuses the concepts and components and when it introduces a new component then it is done in compliance with the STL. Obviously, the scope of Janus is much more specialized than that of the Standard Template Library. Janus shall provide abstractions that are flexible and efficient enough to be usable in (parallel) dynamic mesh problems.

Two points are essential to our approach. The first one is that our abstractions are based on the concept of *static search structures* (see Sect. 2.1). The second point is that we require that a relation between two domains is defined by describing how it *acts* on the data that are defined on the domains (see Sect. 2.3).

2.1   Static Search Structures

Our fundamental concept to represent finite sets and their relations is that of a *static search structure*. We use the following definition of this term [4].

> A *static search structure* is an Abstract Data Type with certain fundamental operations, e.g., initialize, insert, and retrieve. Conceptually, all insertions occur before any retrievals.

What makes static search structure interesting to us is the clear separation between insertions and retrievals. This corresponds to our Two-Phase Property mentioned in the introduction.

2.2   Domains

The concept of a *Domain* formulates requirements how the elements described by a static search structure can be accessed during the retrieval phase.

A *Domain* is a static search structure that describes a set of n elements which are numbered from 0 to $n - 1$. No two of these elements may be the same. Therefore the numbering is a one-to-one mapping. Having this one-to-one correspondence of domain elements and integers plays a fundamental role in our framework. It allows us to apply indirect addressing techniques for an efficient implementation of array operations and relations between domains, see Sect. 2.3 and 3.3 for more details.

The elements described by a domain D are of type D::value_type. A domain must also define an integral type size_type and implement the following non-mutating member functions shown in the following table.

size_type size()	number of elements in the domain
value_type operator[](size_type k)	the kth element of the domain
size_type position(const value_type&)	the inverse of operator[]
bool contains(const value_type& v)	is true iff v belongs to the domain

The test performed by the method contains shall not require more than $O(\ln(n))$ operations where $n = size()$. The ordering implied by the position of the elements of a domain should be understood as a *local index*. This is important since Janus has template classes that represent groups of domains which are not necessarily contained in the same address space.

There are two important refinements of the Domain concept, namely *One Phase Domain* and *Two Phase Domain*, see Fig. 2.

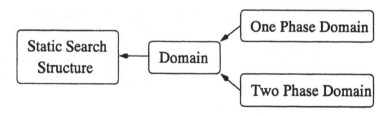

Fig. 2. The domain concept hierarchy. The arrows indicate refinement of a concept.

One Phase Domains correspond to static search structures that do not need insert operations because they can be completely described at initialization time.

Typical examples are rectangular grids and regular subsets thereof. Although these structures are not of primary interest when implementing FEM we mention them to show that regular structures fit also smoothly in the Janus framework. Fig. 3 for example shows how a simple `interval` class can be implemented so that it fulfills the requirements of the One Phase Domain concept. Janus contains so far the classes `grid` and `distributed_grid` to represent regular N-dimensional rectangular grids.

```
class interval { // the domain of the interval [a,b)
  const int a, b;
public:
  // constructors, destructor ... not shown
  typedef size_t size_type; //  types required by the concept Domain
  typedef int    value_type; //
  size_type  size() const  { return b-a; }
  size_type  position(const value_type& x) const { return v-a; }
  value_type operator[](size_type i) const { return a+i; }
  bool contains(const value_type& v) const { return a<=v && v<b; }
};
```

Fig. 3. Class `interval` is a model of the concept one phase domain

The concept of a Two Phase Domain is the more interesting one to represent irregular sets as they occur so frequently in FEM. This concept specifies how elements are inserted in a static search structure and how the phase transition between the insert and retrieval phases is indicated.

For the latter one a method `void freeze()` must be provided. A model of the Two Phase Domain concept must also implement the following methods.

`void insert(const value_type& v)` inserts v into the domain,
`bool frozen() const` returns `true` iff `freeze()` has been called.

In Sect. 3 we describe the template classes `complete_domain` and `distributed_domain` which are the standard models in Janus to represent Two Phase Domains in sequential and MPI contexts, respectively. Note that these classes are also containers in the STL sense which provide iterators accordingly. Since a Domain allows only non-mutating access to its elements these iterators must be non-mutating as well.

2.3 Relations

A primary use of relations in scientific computations is to express communication as generalized matrix-vector multiplication. As an example we consider the nearest neighbor relation on a two-dimensional grid that relates a point (i, j) with the following four points $(i+1, j)$, $(i-1, j)$, $(i, j+1)$, and $(i, j-1)$.

A typical usage of this relation is to describe stencil operations on two-dimensional arrays a and b as they occur in finite difference methods.

```
a[i][j] = b[i+1][j] + b[i-1][j] + b[i][j+1] + b[i][j-1]
```

This is our motivation to consider relations from an algebraic point of view. Before we formulate our requirements we want to emphasize that we have much more complex relations in mind than the simple example shown above. We have to deal with nearest-neighbor and hierarchic relations in potentially hybrid meshes.

Let A, B be two domains, $b : B \to V$ a function and $\oplus$ a binary operation on V. Let $R \subset A \times B$ a relation then we define $R_{\oplus}(b)$ as the function $a : A \to V$ for which $a(i) = \oplus_{(i,j) \in R} b(j)$ hold. We assume that there is a default value 0_V in V that is taken to define $a(i)$ for points i that do not take part in the relation R. Note that this definition of applying a relation to a function can be considered as a matrix vector multiplication where the matrix has the value 1 for pairs (i, j) that belong to the relation R and 0 otherwise. The use of relations as factors in communication products has been inspired by the PROMOTER programming model [12].

The requirements of the concept Relation are based on this definition. A type R that is a model of the concept relation must have a *template member* function apply with the following signature.

```
template<class Domain1, class Domain2> class R<Domain1,Domain2> {
    // ...
    template<class source_it, class target_it>
    void apply(source_it s, target_it t) const;
};
```

The types source_it and target_it must be a random access iterators [7] to types that represent the functions b and a. This makes use of our concept of an array on a domain, see Sect. 3.3. Here we exploit the fact that there is a one-to-one correspondence between domain elements and integers (see Sect. 2.2) which allows to implement operations on arrays by index calculation.

Requiring that the apply method is a template member of the relation type has the advantage that the same relation can be applied to basically any random access container, regardless of its value type. The user can specify the binary relation $\oplus$ used for accumulating elements. The default version of apply which is shown above uses the operator +. In contrast to source_it the iterator target_it must allow mutating access. The size of the containers to which the arguments of the apply method point must be the same as the domains A, B between the relation object exist.

Having an apply method is our fundamental requirement to a relation. Since relations differ in their complexity as domains do we have accordingly the concepts of *One Phase Relation* and *Two Phase Relation*. A finite difference stencil as our first example would be represented by a relation type that is model of the One Phase Relation concept. Irregular relations would be represented by models of Two Phase Relation. For the latter Janus provides two generic relation classes, namely `complete_relation` and `distributed_relation` which are explained in Sect. 3.2.

3   Domain and Relation Template Classes

In this section we describe the major template classes of the Janus library which are shown in Table 1. Since we are primarily interested in irregular mesh applications we only discuss the Two Phase Domains and relations.

Table 1. The domain and relation templates of Janus

Concept	Sequential	Parallel
One Phase Domain	`grid`	`distributed_grid`
One Phase Domain	`polytope`	—
Two Phase Domain	`complete_domain`	`distributed_domain`
Two Phase Relation	`complete_relation`	`distributed_relation`

The main difference between these `complete` and `distributed` classes is that the latter ones are implemented on top of MPI and can be easily used in data parallel programs.

3.1   Details of the Domain Template Classes

The class `complete_domain<T>` is implemented as a sorted array which ensures compact storage and an logarithmic average time to search elements. The implementation uses the `vector` container of the STL and its `lower_bound` binary search algorithm to implement the method `complete_relation::find`. Note that sorting and removing duplicated elements happens when calling the `freeze` method. This means that these operations are almost transparent to the user and are performed only once. We plan to offer a hash based implementation of Two Phase Domains, too.

The template class `distributed_domain` describes a domain that is *distributed* over a fixed group of MPI processes. Locally the class behaves similar to the `complete_domain` template.

The context (process) to which an element of type `value_type` shall be mapped must be specified at insertion time. The communication that is necessary to put the elements in the desired contexts is performed when this container is frozen. This means that the low level details of building and maintaining this distributed data structure are hidden from the user.

Note however, that access to elements is strictly local, that is, this container provides only local and no *parallel* iterators such as the `distributed` containers of the Parallel STL (see Sect. 5). In Janus we strive for the use of iterators in local contexts and offer the concept of Relations (see Sect. 2.3) to access groups of objects.

3.2   Details of the Relation Template Classes

Janus' relation classes complete_relation and distributed_relation are models of the concepts Two Phase Relation. They can explicitly store the ordered pairs of an irregular relation which makes them very attractive for the implementation of finite element programs. Both classes provide not only the member apply(source_it, target_it) as required by the concept Relation 2.3, but also have the member template function

```
template<class matrix_it, class source_it, class target_it>
void apply_matrix(matrix_it m, source_it s, target_it t) const
```

that implements a matrix-vector multiplication $a(i) = \sum_{(i,j) \in R} c(i,j) \cdot b(j)$ where the matrix c is stored in the random access container to which the iterator matrix_it m points. The order in which the matrix is represented must be equal to the one in which the elements of this relation container is stored. For the latter the Compressed Row Storage (CRS) format is used. As in the case of the class complete_domain the final storage scheme is build transparently to the user when the freeze method is called.

It might be surprising that the code for the matrix-vector multiplication is contained in the relation class. However, this is where it belongs to since an efficient implementation requires knowledge of the internal structures of the relation. Using member templates is a clean way to make this knowledge general accessible.

3.3   Arrays and Sparse Matrices in Janus

As in the POOMA [9] framework we consider an *array* as a function that maps from a input domain to a range of outputs. Since our domains are linearly ordered we can fall back on one-dimensional arrays to represent general arrays on them. A value in a one-dimensional array and a particular element of a domain are associated if they have the same position in their respective structures. Since random access is supported only local within each context this implies that an array is distributed according to the underlying domain.

This simple and intuitive interface makes it unnecessary to provide an own array class in Janus. We decided to use the std::valarray template which offers a variety of overloaded operators and mathematical functions and several ways to extract subsets of arrays.

We also have a template class sparse_matrix<Relation,T> that is parameterized over the values of the matrix and the type of the underlying relation. This class sores the entries of the matrix and uses the apply_matrix member template method of the underlying relation to perform the matrix-vector multiplication (see Sect. 3.2. This approach allows sharing the same relation (i.e. sparseness pattern) for different sparse_matrix objects.

4   Application to Parallel FEM

We have written a simple program that can read and refine two-dimensional triangulations, build the element matrices of the Laplace operator for linear triangle
elements, and solve the assembled sparse algebraic system with the conjugate
gradient method. Diagonal scaling is used as preconditioner. The triangulations
were generated and displayed with the TRIANGLE tool set developed by J.R.
Shewchuk [13]. The partitions of the triangulations were generated with the
METIS tool set developed by G. Karypis [14].

We used up to 128 processors of the Linux PC cluster of the Real World
Computing Partnership [15]. As C++ compiler g++ with the optimization op-
tion "-O3" has been used. We used MPICH on the PM [16] communication
library that supports the Myrinet network of the RWC cluster. The problem
mesh consisted of 131,072 triangles and 66,049 nodes. The assembled system
matrix had 460,289 entries. In the benchmark only the time for the preconditioned conjugate gradient method was measured. We compared a pure sequential
version that does not include any call to parallel communication primitives and
a parallel version that was run on 1, 2, 4, ..., 128 processors. The runtime the
preconditioned conjugate gradient method is shown in Table 5. This table also
includes the absolute speedup defined as the ratio of the runtime of sequential
version to a particular version.

The highest speedup is 16 when using 32 nodes. Here the parallel version
is on one processor almost two times slower than the sequential version. This
makes the result on the PC cluster look much worse.

5   Related Work

The Parallel Standard Template Library (PSTL) of HPC++ is a parallel extension of the C++ Standard Template Library [8]. The PSTL provides distributed versions of the STL container classes (vector, list, map, ...). These
new containers have the prefix distributed_ in their name and provide beside
the standard iterator interface *parallel iterators* which are a generalization of
global pointers. The PSTL also includes a multidimensional array class which
supports element access via standard array indexing and parallel iterators.

Like their sequential counterparts the distributed containers support insertions and deletions of elements. This is different to our container classes which
only supports insertions. Although this sounds like a major restriction it is not
since our domains are static search structures whose usage conforms with the
requirements of finite sets and relations as they occur in numerical methods.

POOMA [9] is a application-oriented framework that is organized into several
abstraction layers and is implemented in ISO-C++. On the lowest level POOMA
provides generic containers to model domains and arrays. This is the level on
which POOMA and Janus can be compared.

POOMA offers several template classes to represent different kinds of sub-
grids of Z^n. It is not clear how complicated spatial structures such as finite
element meshes can be efficiently represented within this framework.

# PE	time (seconds)	speedup
sequential	92.9899	1.0
1	180.956	0.514
2	80.4036	1.157
4	36.6468	2.537
8	17.8723	5.203
16	9.03195	10.296
32	5.51059	16.875
64	7.42557	12.523
128	22.2499	4.179

Fig. 4. Absolute speedup of the conjugate gradient method on the RWC Linux PC Cluster. All times are in seconds. The right figure shows a two-dimensional triangulation of a rectangle with hole that is partitioned into 16 pieces.

POOMA takes a sophisticated two layered approach to represent arrays on these domains. On the lower level there is the concept of an *Engine* which provides a common interface for randomly accessing and changing data. How the data are stored and indexed is hidden by the particular engine type. On the higher level there is the template class Array that describes a logical rectangular, N-dimensional table of elements. There is no performance penalty in this indirection because the gap between the array and its engine template parameter is resolved at compile time.

As explained in Sect. 3.3 we took a different approach for representing arrays in Janus. We experienced that the explicit decoupling of domain and array types reduced the compilation time considerably. Long compilation times and big executables are still a problem when using template libraries in particular when they contain many closely intertwined template classes and functions.

The Matrix Template Library (MTL) [10] is a very consequent way of applying generic programming in the field of high performance computing. It offers a broad variety of matrix container concepts and classes–much more than Janus. The main difference of these two libraries is there focus. Janus concentrates on spatial patterns of numeric applications and considers matrices as functions on these patterns (relations). It would be interesting for Janus and the MTL if their relation and matrix classes could be easily be combined.

6   Conclusions and Future Work

In this paper we have presented the C++ template library Janus which offers container classes that are useful for the compact and efficient representation of finite sets and relations. Our containers are models of the Domain and Relation concepts that are based on the idea of static search structures and STL concepts. One key to understand our approach is to realize that a static search structure is usually flexible enough to capture the dynamicity of finite element meshes and sparse matrix graphs. Another fundamental property of our approach is that we consequently use relation types to express communication. Technically, the phase transition of static search structures is implemented within a `freeze` method. Necessary, low level operations like sorting, sending and receiving can be performed when calling `freeze` so that they are basically transparent to the user.

Different to many other C++ libraries for scientific programming Janus provides no general multi-dimensional arrays class. Since domains have random access we can use standard arrays like `valarray` to represent functions on domains. This is more flexible since our domain classes are not restricted to rectangular, grid-like structures.

One constraint in the design of our FEM program was that it should always be possible to built a purely sequential version. Using Janus it is easy to switch between the two implementation modes. Basically the user must replace all classes with prefix `distributed_` against their counterparts with the prefix `complete_` This change can be localized when using typedefs instead of the "raw" template names.

As an example we used Janus for the parallel implementation of a two-dimensional FEM problem. On a Linux PC cluster we achieved maximal absolute speedup of 16 when using 32 processors. The main reason for this is that we our sparse matrix multiplication isn't implemented optimally yet.

An often asked question is whether we should have a *thaw* method that would allow to switch from the retrieval phase to the modification phase again. We do think that both from a conceptual and implementation point of view it is better to avoid thawing of frozen objects. Rather we recommend creating new object and perform the necessary modifications while inserting into them.

Our classes do currently not support multithreading environments directly. However, using OpenMP [17] this could be done within in the domain and relation classes, mostly transparent to the user.

We also have the idea to implement our framework of Domain and Relation concept in Java when direct support for parameterized types is available.

The first author especially wants to thank his colleague Hans-Werner Pohl for the fruitful discussions about the ideas of Janus and for writing the `polytope` domain template class.

References

1. P. Bastian, K. Birken, S. Johansen K., Lang, N. Neuss, H. Rentz-Reichert, and C. Wieners. *UG – a Flexible Software Toolbox for Solving Partial Differential Equations*. Institut für Computeranwendungen III, Universität Stuttgart, Pfaffen-waldring 27, 70569 Stuttgart. http://www.ica3.uni-stuttgart.de.
2. M. Griebel and G. Zumbusch. Hash-Storage Techniques for Adaptive Multilevel Solvers and their Domain Decomposition Parallelization. In Mandel, J. and Farhat, C. and Cai, X.-C., editor, *Proceedings of Domain Decomposition Methods 10*, volume 218 of *Contemporary Mathematics*, pages 279–286. AMS, 1998.

3. J. Gerlach, M. Sato, and Y. Ishikawa. Janus: A C++ Template Library for Parallel Dynamic Mesh Applications. In *Proceedings of the Second International Symposium of Computing in Object-Oriented Parallel Environments ISCOPE 99*, volume 1505 of *Lecture Notes in Computer Science*, pages 215–222, Santa Fe, New Mexico, USA, December 1998. Springer-Verlag.
4. Douglas C. Schmidt. *Users Guide to gperf.* Free Software Foundation. http://www.gnu.org/manual/gperf-2.7/gperf.html.
5. D. R. Musser and A. A. Stepanov. Generic Programming. In *First International Joint Conference of ISSAC-88 and AAECC-6*, volume 358 of *Lecture Notes in Computer Science*, pages 13–25. Springer, June 1988.
6. D. Musser. *Generic Programming.* http://www.cs.rpi.edu/~musser/gp/index.html.
7. Silicon Graphics Computer Systems, Inc. *Standard Template Library Programmer's Guide.* http://www.sgi.com/Technology/STL.
8. E. Johnson and D. Gannon. Programming with the HPC++ Parallel Standard Template Library. In *SIAM Conference on Parallel Processing for Scientific Computing*, Minneapolis, Minnesota, March 1997.
9. Steve Karmesin, James Crotinger, Julian Cummings, Scott Haney, William Humphrey, John Reynders, Stephen Smith, and Timothy J. Williams. Array Design and Expression Evaluation in POOMA II. In *Proceedings of the Second International Symposium of Computing in Object-Oriented Parallel Environments ISCOPE 99*, volume 1505 of *Lecture Notes in Computer Science*, pages 231–238, Santa Fe, New Mexico, USA, December 1998. Springer-Verlag.
10. J. G. Siek and A. Lumsdaine. The Matrix Template Library: A Generic Programming Approach to High Performance Numerical Algebra. In *Proceedings of the Second International Symposium of Computing in Object-Oriented Parallel Environments ISCOPE 99*, volume 1505 of *Lecture Notes in Computer Science*, pages 59–70, Santa Fe, New Mexico, USA, December 1998. Springer-Verlag.
11. B. Stroustrup. *The C++ Programming Language, Third Edition.* Addison-Wesley, 1998.
12. W.K. Giloi, M. Kessler, and A Schramm. PROMOTER : A High Level Object-Parallel Programming Language. In *Proceedings of International Conference on High Performance Computing*, New Delhi, India, December 1995.
13. J. R. Shewchuk. *A Two-Dimensional Quality Mesh Generator and Delaunay Triangulator.* http://www.cs.cmu.edu/~quake/triangle.html.
14. G. Karypis. *Metis a Family of Multilevel Partitioning Algorithms.* http://www-users.cs.umn.edu/~karypis/metis/main.shtml.
15. Yutaka Ishikawa, Hiroshi Tezuka, Atsuhi Hori, Shinji Sumimoto, Toshiyuki Takahashi, Francis O'Carroll, and Hiroshi Harada. RWC PC Cluster II and SCore Cluster System Software – High Performance Linux Cluster. In *Proceedings of the 5th Annual Linux Expo*, pages 55 – 62, 1999.
16. Hiroshi Tezuka, Atsushi Hori, Yutaka Ishikawa, and Mitsuhisa Sato. PM: An Operating System Coordinated High Performance Communication Library. In Bob Hertzberger Peter Sloot, editor, *High-Performance Computing and Networking*, volume 1225 of *Lecture Notes in Computer Science*, pages 708–717. Springer, April 1997.
17. The OpenMP Consortium, www.openmp.org. *OpenMP: A Proposed Industry Standard API for Shared Memory Programming*, October 1997. http://www.openmp.org/mp-documents/paper/paper.ps.

Generic Graph Algorithms
for Sparse Matrix Ordering*

Lie-Quan Lee, Jeremy G. Siek, and Andrew Lumsdaine

University of Notre Dame, Notre Dame, IN, USA

Abstract. Fill-reducing sparse matrix orderings have been a topic of active research for many years. Although most such algorithms are developed and analyzed within a graph-theoretical framework, for reasons of performance the corresponding implementations are typically realized with programming languages devoid of language features necessary to explicitly represent graph abstractions. Recently, generic programming has emerged as a programming paradigm capable of providing high levels of performance in the presence of programming abstractions. In this paper we present an implementation of the Minimum Degree ordering algorithm using the newly-developed Generic Graph Component Library. Experimental comparisons show that, despite our heavy use of abstractions, our implementation has performance indistinguishable from that of a widely used Fortran implementation.

1 Introduction

Computations with symmetric positive definite sparse matrices are a common and important task in scientific computing. For efficient matrix factorization and linear system solution, the ordering of the equations plays an important role. Because Gaussian elimination (without numerical pivoting) of symmetric positive definite systems is stable, such systems can be ordered before factorization takes place based only on the structure of the sparse matrix. Unfortunately, determining the optimal ordering (in the sense of minimizing fill-in) is an NP-complete problem [1], so greedy heuristic algorithms are typically used instead.

Algorithms for sparse matrix ordering have been an active research topic for many years. These algorithms are typically developed in graph-theoretical terms, while the most widely used implementations are coded in Fortran 77. Since Fortran 77 supports no abstract data types other than arrays, the graph abstractions used to develop and describe the ordering algorithms must be discarded in the actual implementation. Although graph algorithms are well-developed and widely-implemented in higher-level languages such as C or C++, performance concerns (which are often paramount in scientific computing) have continued to restrict implementations of sparse matrix ordering algorithms to Fortran.

Efforts to develop sparse matrix orderings with modern programming techniques include [2] and [3]. These were based on an object-oriented, rather than

* This work was supported by NSF grants ASC94-22380 and CCR95-02710.

S. Matsuoka et al. (Eds.): ISCOPE'99, LNCS 1732, pp. 120–129, 1999.
© Springer-Verlag Berlin Heidelberg 1999

generic, programming paradigm and although they were well programmed, the reported performance was still a factor of 4-5 slower than well-established Fortran 77 implementations.

The recently introduced programming paradigm known as *generic programming* [4,5] has demonstrated that abstraction and performance are not necessarily mutually exclusive. One example of a graph library that incorporates the generic programming paradigm is the recently developed Generic Graph Component Library (GGCL) [6]. In this paper we present an implementation of the minimum degree algorithm for sparse matrix ordering using the GGCL. Although the implementation uses powerful graph abstractions, its performance is indistinguishable from that of one of the most widely used Fortran 77 codes.

2   Generic Programming

Generic programming is a powerful new paradigm for software development, particularly for the development of (and use of) component libraries. The most visible (and perhaps most important) popular example of generic programming is the celebrated Standard Template Library (STL) [7]. The fundamental principle of generic programming is to separate algorithms from the concrete data structures on which they operate, based on abstract concepts of the underlying problem domain. That is, in a generic library, algorithms do not manipulate concrete data structures directly, but instead operate on abstract interfaces defined for entire equivalence classes of data structures. A single generic algorithm can thus be applied to any particular data structure that conforms to the requirements of its equivalence class.

In STL the data structures are *containers* such as vectors and linked lists and *iterators* form the abstract interface between *algorithms* and containers. Each STL algorithm is written in terms of the iterator interface and as a result each algorithm can operate with any of the STL containers. In addition, many of the STL algorithms are parameterized not only on the type of iterator being accessed, but on the type of operation that is applied during the traversal of a container as well. For example, the `transform()` algorithm has a parameter for a `UnaryOperator` *function object*. Finally, STL contains classes known as *adaptors* that are used to modify underlying class interfaces.

The generic programming approach to software development can provide tremendous benefits to such aspects of software quality as functionality, reliability, usability, maintainability, portability, and efficiency. The last point, efficiency, is of particular (and sometimes paramount) concern in scientific applications. Performance is often of such importance to scientific applications that other aspects of software quality may be deliberately sacrificed if elegant programming abstractions and high performance cannot be simultaneously achieved. Until quite recently, the common wisdom has been that high levels of abstraction and high levels of performance were, *per se*, mutually exclusive. However, beginning with STL for general-purpose programming, and continuing with the Matrix Template Library (MTL) [5] for linear algebra, it has been clearly demonstrated that abstraction does not necessarily preclude performance. In fact, MTL provides performance equivalent to that of highly-optimized vendor-tuned math libraries.

3   The Generic Graph Component Library

The Generic Graph Component Library (GGCL) is a collection of high-performance graph algorithms and data structures, written in C++ using the generic programming style. Although the domain of graphs and graph algorithms is a natural one for the application of generic programming, there are important (and fundamental) differences between the types of algorithms and data structures in STL and the types of algorithms and data structures in a generic graph library.

To describe the graph interface of GGCL we use generic programming terminology from the SGI STL [4]. In that parlance, the set of requirements on a template parameter for a generic algorithm or data structure is called a *concept*. The various classes that fulfill the requirements of a concept are said to be *models* of the concept. Concepts can extend other concepts, which is referred to as *refinement*. We use a bold sans serif font for concept identifiers.

3.1   Graph Concepts

The graph interface used by GGCL can be derived directly from the formal definition of a graph [8]. A graph G is a pair (V,E), where V is a finite set and E is a binary relation on V. V is called a *vertex set* whose elements are called *vertices*. E is called an *edge set* whose elements are called *edges*. An edge is an ordered or unordered pair (u,v) where $u,v \in V$. If (u,v) is and edge in graph G, then vertex v is *adjacent* to vertex u. Edge (u,v) is an *out-edge* of vertex u and an *in-edge* of vertex v. In a *directed* graph edges are ordered pairs while in a *undirected* graph edges are unordered pairs. In a *directed* graph an edge (u,v) leaves from the *source* vertex u to the *target* vertex v.

The three main concepts necessary to define our graph are Graph, Vertex, and Edge. The abstract iterator interface used by STL is not sufficiently rich to encompass the numerous ways that graph algorithms may compute with a graph. Instead, we formulate an abstract interface, based on Visitor and Decorator concepts, that serves the same purpose for graphs that iterators do for basic containers. These two concepts are similar in spirit to the "Gang of Four" [9] patterns of the same name, however the implementation techniques used are based on static polymorphism and mixins [10] instead of dynamic polymorphism. Fig. 1 depicts the analogy between the STL and the GGCL.

Graph: The Graph concept merely contains a set of vertices and a set of edges and a tag to specify whether it is a directed graph or an undirected graph. The only requirement is the *vertex set* be a model of Container and its value_type a model of Vertex. The *edge set* must be a model of Container and its value_type a model of Edge.

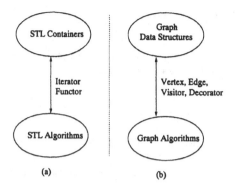

Fig. 1. The analogy between the STL and the GGCL.

Vertex: The Vertex concept provides access to the adjacent vertices, the out-edges of the vertex and optionally the in-edges.

Edge: An Edge is a pair of vertices, one is the *source* vertex and the other is the *target* vertex. In the unordered case it is just assumed that the position of the *source* and *target* vertices are interchangeable.

Decorator: As mentioned in the introduction, we would like to have a generic mechanism for accessing vertex and edge properties (such as color and weight) from within an algorithm. Such an access method is necessary because of the many ways in which the properties may be stored, and the many ways in which access to that storage may be implemented. We give the name Decorator to the concept for this generic access method. The implementation of graph Decorators is similar in spirit to the "Gang of Four" (GoF) decorator pattern [9]. In many respects a Decorator is similar to a functor, or function object. However, we use operator[] instead of operator() since it results in better match for the commonly used graph algorithm notations.

Visitor: In the same way that function objects are used to make STL algorithms more flexible, we can use functor-like objects to make the graph algorithms more flexible. We use the name Visitor for this concept, since we are basically just using a template version of the well known GoF visitor pattern [9]. Our Visitor is somewhat more complex than a function object, since there are several well defined entry points at which the user may want to introduce a call-back.

The Decorator and Visitor concepts are used in the GGCL graph algorithm interfaces to allow for maximum flexibility. Below is the prototype for the GGCL depth first search algorithm, which includes parameters for both a Decorator and a Visitor object. There are two overloaded versions of the interface, one of which has a default ColorDecorator. The default decorator accesses the color property directly from the graph vertex. The STL overloads generic functions in this way as well. For example, there are two overloaded versions of the lower_bound() algorithm. One uses operator< by default and the other takes a BinaryOperator functor argument.

```
template <class Graph, class Visitor>
void dfs(Graph& G, Visitor visit);

template <class Graph, class Visitor, class ColorDecorator>
void dfs(Graph& G, Visitor visit, ColorDecorator color);
```

3.2   Generic Graph Algorithms

With the abstract graph interface defined, generic graph algorithms can be written solely in terms of the graph interface. The algorithms do not make any assumptions about the actual underlying graph data structure.

```
template <class Vertex, class QType, class Visitor>
void generalized_BFS(Vertex s, QType& Q, Visitor visitor)
{
  vertex_traits<Vertex>::edgelist_type::iterator ei;
  typedef typename vertex_traits<Vertex>::edge_type Edge;
  visitor.start(s);
  Q.push(s);
  while (! Q.empty()) {
    Vertex u = Q.front();
    Q.pop();
    visitor.discover(u);
    for (ei = out_edges(u).begin();
         ei != out_edges(u).end(); ++ei) {
      Edge e = *ei;
      if (visitor.process(e))
        Q.push(target(e));
    }
    visitor.finish(u);
  }
}
```

Fig. 2. The generalized Breadth First Search algorithm.

The Breadth First Search (BFS) algorithm, as an example, is shown in Fig. 3.2. In this algorithm we use the expression out_edges(u) to access the ContainerRef of edges leaving vertex u. We can then use the iterators of this Container to access each of the edges. In this algorithm, the Visitor is used to abstract the kind of operation performed on each edge as it is discovered. The algorithm also inserts each discovered vertex onto Q. The vertex is accessed through e.target().

The concise implementation of algorithms is enabled by the genericity of the GGCL algorithms, allowing us to exploit the reuse that is inherent in these graph algorithms in a concrete fashion.

4 Sparse Matrix Ordering

The process for solving a sparse symmetric positive definite linear system, $Ax = b$, can be divided into four stages as follows:

Ordering: Find a permutation P of matrix A,
Symbolic factorization: Set up a data structure for Cholesky factor L,
Numerical factorization: Decompose PAP^T into LL^T,
Triangular system solution: Solve $(LL^T)Px = Pb$ for x.

Because the choice of permutation P will directly determine the number of fill-in elements (elements present in the non-zero structure of L that are not present in the non-zero structure of A), the ordering has a significant impact on the memory and computational requirements for the latter stages. However, finding the optimal ordering for A (in the sense of minimizing fill-in) has been proven to be NP-complete [1] requiring that heuristics be used for all but simple (or specially structured) cases.

Developing algorithms for high-quality orderings has been an active research topic for many years. Most ordering algorithms in wide use are based on a greedy approach such that the ordering is chosen to minimize some quantity at each step of a simulated n-step symmetric Gaussian elimination process. Algorithms using such an approach are typically distinguished by their minimization criteria [11].

4.1 Graph Models

In 1961, Parter introduced the graph model of symmetric Gaussian elimination [12]. A sequence of elimination graphs represent a sequence of Gaussian elimination steps. The initial Elimination graph is the original graph for matrix A. The elimination graph at the kth step is obtained by adding edges between adjacent vertices of the current eliminated vertex to form a clique and removing the eliminated vertex and its edges.

In graph terms, the basic ordering process used by most greedy algorithms is as follows:

1. *Start:* Construct undirected graph G^0 corresponding to matrix A
2. *Iterate:* For $k = 0, 1, \ldots$, until $G^k = \emptyset$ do:
   - Choose a vertex v^k from G^k according to some criterion
   - Eliminate v^k from G^k to form G^{k+1}

The sequence of vertices $\{v^0, v^1, \ldots\}$ selected by the algorithm defines the ordering.

One of the most important examples of such an algorithm is the *Minimum Degree* algorithm. At each step the minimum degree algorithm chooses the vertex with minimum degree in the corresponding graph as v^k. A number of enhancements to the basic minimum degree algorithm have been developed, such as the use of a quotient graph representation, mass elimination, incomplete degree update, multiple elimination, and external degree. See [13] for a historical survey of the minimum degree algorithm. Many of these enhancements, although initially proposed for the minimum degree algorithm, can be applied to other greedy approaches as well. Other greedy approaches differ from minimum degree by the choice of minimization criteria for choosing new vertices. For example, to accelerate one of the primary bottlenecks of the ordering process, the *Approximate Minimum Degree* (AMD) algorithm uses an estimate of the degree (or external degree) of a vertex [14]. The *Minimum Deficiency* class of algorithms instead choose the vertex that would create the minimum number of fill-in elements. A nice comparison of many of these different approaches can be found in [11].

5   Implementation

Our GGCL-based implementation of MMD closely follows the algorithmic descriptions of MMD given, e.g., in [13,15]. The implementation presently includes the enhancements for mass elimination, incomplete degree update, multiple elimination, and external degree. In addition, we use a quotient-graph representation. Some particular details of our implementation are given below.

Prototype Our algorithm is prototyped as follows:

```
template<class Graph, class RandomAccessContainer,
         class Decorator>
void mmd(Graph& G, RandomAccessContainer& Permutation,
    RandomAccessContainer& InversePermutation,
    Decorator SuperNodeSize, int delta = 0)
```

The parameters are used in the following ways:

G (input/output) is the graph representing the matrix A to be ordered on input. On output, G contains the results of the ordered elimination process. May be used subsequently by symbolic factorization.

Permutation, InversePermutation (output) respectively contain the permutation and inverse permutation produced by the algorithm.

SuperNodeSize (output) contains the size of supernodes or supernode representative node produced by the the algorithm. May be used subsequently by symbolic factorization.

delta (input) controls multiple elimination.

Abstract Graph Representation Our minimum degree algorithm is expressed only in terms of the GGCL abstract graph interface. Thus, any underlying concrete representation that models the GGCL Graph concept can be used. However, not all concrete representations will provide the same levels of performance. A high performance representation is described next.

Concrete Graph Representation We use an adjacency list representation within the GGCL framework. In particular the graph is based on a templated "vector of vectors." The vector container used is an adaptor class built on top the STL vector class. Particular characteristics of this adaptor include the following:

- Erasing elements does not shrink the associated memory. Adding new elements after erasing will not need to allocate additional memory.
- Additional memory is allocated efficiently on demand when new elements are added (doubling the capacity every time it is increased). This property comes from STL vector.

We note that this representation is similar to that used in Liu's implementation, with some important differences due to dynamic memory allocation. With the dynamic memory allocation we do not need to over-write portions of the graph that have been eliminated, allowing for a more efficient graph traversal. More importantly, information about the elimination graph is preserved allowing for trivial symbolic factorization. Since symbolic factorization can be an expensive part of the entire solution process, improving its performance can result in significant computational savings.

The overhead of dynamic memory allocation could conceivably compromise performance in some cases. However, in practice, memory allocation overhead does not contribute significantly to run-time for our MMD implementation. Finally, with our approach, somewhat more total memory may be required for graph representation. In the context of the entire sparse matrix solution process this is not an important issue because the memory used for the graph during ordering can be returned to the system for use in subsequent stages (which would use more memory than even the dynamically-allocated graph at any rate).

6 Experimental Results

The performance of our implementation was tested with selected matrices from the Harwell-Boeing collection [16] and the University of Florida's sparse matrix collection [17], as well as with locally-generated discretized Laplacians.

For our tests, we compare the execution time of our implementation of MMD against that of the equivalent SPARSPAK algorithm (GENMMD). The tests were run on a Sun SPARC Station U-30 having a 300MHz UltraSPARC-II processor, 256MB RAM, and Solaris 2.6. The GENMMD code was compiled with Solaris F77 4.2 with optimizing flags -fast -xdepend -xtarget=ultra2 -xarch=v8plus -x04 -stackvar -xsafe=mem. The C++ code was compiled with Kuck and Associates KCC version 3.3e using aggressive optimization for the C++ front-end. The back-end compiler was Solaris cc version 4.2, using optimizations basically equivalent to those given above for the Fortran compiler.

Table 1 gives the performance results. For each case, our implementation and GENMMD produced identical orderings. Note that the performance of our implementation is essentially equal to that of the Fortran implementation and even surpasses the Fortran implementation in a few cases.

Matrix	n	nnz	GENMMD	GGCL
BCSPWR09	1723	2394	0.00728841	0.007807
BCSPWR10	5300	8271	0.0306503	0.033222
BCSSTK15	3948	56934	0.13866	0.142741
BCSSTK18	11948	68571	0.251257	0.258589
BCSSTK21	3600	11500	0.0339959	0.039638
BCSSTK23	3134	21022	0.150273	0.146198
BCSSTK24	3562	78174	0.0305037	0.031361
BCSSTK26	1922	14207	0.0262676	0.026178
BCSSTK27	1224	27451	0.00987525	0.010078
BCSSTK28	4410	107307	0.0435296	0.044423
BCSSTK29	13992	302748	0.344164	0.352947
BCSSTK31	35588	572914	0.842505	0.884734
BCSSTK35	30237	709963	0.532725	0.580499
BCSSTK36	23052	560044	0.302156	0.333226
BCSSTK37	25503	557737	0.347472	0.369738
CRYSTK02	13965	477309	0.239564	0.250633
CRYSTK03	24696	863241	0.455818	0.480006
CRYSTM03	24696	279537	0.293619	0.366581
CT20STIF	52329	1323067	1.59866	1.59809
LA2D64	4096	8064	0.022337	0.028669
LA2D128	16384	32512	0.0916937	0.119037
LA3D16	4096	11520	0.0765908	0.077862
LA3D32	32768	95232	0.87223	0.882814
PWT	36519	144794	0.312136	0.383882
SHUTTLE_EDDY	10429	46585	0.0546211	0.066164
NASASRB	54870	1311227	1.34424	1.30256

Table 1. Test matrices and ordering time in seconds, for GENMMD (Fortran) and GGCL (C++) implementations of minimum degree ordering. Also shown are the matrix order (n) and the number of off-diagonal non-zero elements (nnz).

7   Future Work

The work reported here only scratches the surface of what is possible using GGCL for sparse matrix orderings (or more generally, using generic programming for sparse matrix computations). The highly modular nature of generic programs makes the implementations of entire classes of algorithms possible. For instance, a generalized greedy ordering algorithm (currently being developed) will enable the immediate implementation of most (if not all) of the greedy algorithms related to MMD (e.g., minimum deficiency). We are also working to develop super-node based sparse matrices as part of the Matrix Template Library and in fact to develop all of the necessary infrastructure for a complete generic high-performance sparse matrix package. Future work will extend these approaches from the symmetric positive definite case to the general case.

References

1. M Yannanakis. Computing the minimum fill-in is NP-complete. *SIAM Journal of Algebraic and Discrete Methods*, 1981.
2. Kaixiang Zhong. A sparse matrix package using the standard template library. Master's thesis, University of Notre Dame, 1996.
3. Gary Kumfert and Alex Pothen. An object-oriented collection of minimum degree algorithms. In *Computing in Object-Oriented Parallel Environments*, pages 95–106, 1998.
4. Matthew H. Austern. *Generic Programming and the STL*. Addison Wesley Longman, Inc, October 1998.
5. Jeremy G. Siek and Andrew Lumsdaine. The matrix template library: A generic programming approach to high performance numerical linear algebra. In Denis Carmel, Rodney R. Oldhhoeft, and Marydell Tholburn, editors, *Computing in Object-Oriented Parallel Environments*, pages 59–70, 1998.
6. Lie-Quan Lee, Jeremy G. Siek, and Andrew Lumsdaine. The generic graph component library. In *OOPSLA'99*, 1999. Accepted.
7. Meng Lee and Alexander Stepanov. The standard template library. Technical report, HP Laboratories, February 1995.
8. Thomas H. Cormen, Charles E. Leiserson, and Ronald L. Rivest. *Introduction to Algorithms*. The MIT Press, 1990.
9. Erich Gamma, Richard Helm, Ralph Johnson, and John Vlissides. *Design Patterns: Elements of Reusable Object-Oriented Software*. Addiaon Wesley Publishing Company, October 1994.
10. Yannis Samaragdakis and Don Batory. Implementing layered designs with mixin layers. In *The Europe Conference on Object-Oriented Programming*, 1998.
11. Esmond G. Ng amd Padma Raghavan. Performance of greedy ordering heuristics for sparse Cholesky factorization. *SIAM Journal on Matrix Analysis and Applications*, To appear.
12. S. Parter. The use of planar graph in Gaussian elimination. *SIAM Review*, 3:364–369, 1961.
13. Alan George and Joseph W. H. Liu. The evolution of the minimum degree ordering algorithm. *SIAM Review*, 31(1):1–19, March 1989.
14. Patrick Amestoy, Timothy A. Davis, and Iain S. Duff. An approximation minimum degree ordering algorithm. *SIAM J. Matrix Analysis and Applications*, 17(4):886–905, 1996.
15. Joseph W. H. Liu. Modification of the minimum-degree algorithm by multiple elimination. *ACM Transaction on Mathematical Software*, 11(2):141–153, 1985.
16. Roger G. Grimes, John G. Lewis, and Iain S. Duff. User's guide for the harwell-boeing sparse matrix collection. User's Manual Release 1, Boeing Computer Services, Seattle, WA, October 1992.
17. University of Florida sparse matrix collection.
   http://www-pub.cise.ufl.edu/~davis/sparse/.

Using Object-Oriented Techniques for Realizing Parallel Architectural Skeletons

Dhrubajyoti Goswami, Ajit Singh, and Bruno R. Preiss

University of Waterloo, Waterloo, Ontario, Canada

Abstract. The concept of design patterns has recently emerged as a new paradigm in the context of object-oriented design methodology. Similar ideas are being explored in other areas of computing. In the parallel computing domain, design patterns describe recurring parallel computing problems and their solution strategies. Starting with the late 1980's, several pattern-based systems have been built for facilitating parallel application development. However, most of these systems use patterns in ad hoc manners, thus lacking a generic or standard model for using and intermixing different patterns. This substantially hampers the usability of such systems. Lack of flexibility and extensibility are some of the other major concerns associated with most of these systems. In this paper, we propose a generic (i.e., pattern- and application-independent) model for realizing and using parallel design patterns. The term *architectural skeleton* is used to represent the application independent, reusable set of attributes associated with a pattern. The model can provide most of the functionalities of low level message passing libraries, such as PVM or MPI, plus the benefits of the patterns. This results in tremendous flexibility to the user. It turns out that the model is an ideal candidate for object-oriented style of design and implementation. It is currently implemented as a C++ template-library without requiring any language extension. The generic model, together with the object-oriented and library-based approach, facilitates extensibility.

1 Introduction

In the context of object-oriented design methodologies, *design patterns* [1] are used to specify solution strategies for solving recurring design problems in systematic and general ways. Similarly, in the parallel computing domain, design patterns specify recurring parallel computational patterns and their solution strategies. Examples of such recurring patterns are: static and dynamic replication, divide and conquer, data parallel pattern with various topologies, compositional framework for irregularly-structured control-parallel computation, systolic array, singleton pattern for single-process (i.e., sequential) computation.

Starting with the late 1980s, several pattern-based systems have been built, e.g. Frameworks and Enterprise [2], CODE and HeNCE [3], DPnDP [4], for facilitating parallel application development. In a separate but similar thread of work, a group of researchers started exploring parallel patterns as high level functional constructs. The term *algorithmic skeleton* was introduced [5] to specify

S. Matsuoka et al. (Eds.): ISCOPE'99, LNCS 1732, pp. 130–141, 1999.

higher-order functions with specific implementations tailored to particular parallel architectures. The algorithmic skeleton research [6] concentrates on various functional and logic programming languages for abstracting and representing recurring parallel patterns.

Some other approaches have taken the path of building a new language to support program composition using skeletons [7], while others are based on the extension of existing languages, e.g. Frameworks and Enterprise [2].

Most of the pattern-based systems mentioned previously support only a limited set of patterns in ad hoc ways. There is no generic or canonical model of a pattern. This substantially hampers the usability of such systems. Besides usability, there are two other very important aspects: flexibility and extensibility. Most of the systems are hard-coded with a fixed set of patterns, and there is no clear way to add new patterns to the system when required (i.e., lack of extensibility). Besides, if a certain desired parallel structure is not supported by a design-pattern-based system, there is often no alternative but to entirely abandon the idea of using the particular system (i.e., lack of flexibility). A detailed discussion regarding the desirable characteristics and the shortcomings of some of these systems can be found in [2].

This paper defines an *architectural skeleton* as a generic building-block which captures a parallel design pattern in an application-independent manner. Once the required structural and behavioral parameters are assigned to an architectural skeleton, it results in an *abstract parallel computing module* (abbreviated as an *abstract module*). User subsequently attaches application code to an abstract module, which results in a *concrete parallel computing module* (abbreviated as a *concrete module* or simply a *module*). A parallel application consists of one or more mutually interacting, concrete modules.

The model supports top-down hierarchical refinements such that a module, abstract or concrete, can be a composition of other modules. Consequently, every parallel application can be structured in a top-down hierarchical fashion, starting with the root of the hierarchy.

At the beginning of this research, we were not committed to any particular design methodology or implementation style. However, as the work proceeded, the model with the desired capabilities turned out to be an ideal candidate for object-oriented style of design and implementation. Presently, it has been fully implemented using industry-standard C++. The generic model, together with the object-oriented and library based approach, facilitates extensibility. That is, new patterns can be added to the library without requiring any overall change to the existing system.

A pattern can support low- or high-level communication and synchronization protocols. At the same time, multiple patterns can work together based on a generic scheme. Support for hierarchical design and low-level as well as high-level protocols provide the additional flexibility, not found in the existing parallel systems that aim to support design patterns.

2   The Architectural Skeleton Based Model

We start this section with an example. The objective is to introduce the various concepts of the model and its implementation, before elaborating them in detail. The same example will be used at various points throughout the rest of the discussion. Words within *italics* are terms with special meanings in the context of the model and are defined subsequently.

2.1   An Example

The following example illustrates a master-worker application, where the Master *module* produces a succession of image frames and sends them to dynamically replicated Worker *modules* for processing. The Master *module* is implemented as an extension of the replication *skeleton* which supports arbitrary degree of replication that is dynamically controlled at run-time, depending on the workload and the number of available processors. Each replicated Worker *module* extends a singleton *skeleton*, which is used for single-process computation. Each replicated Worker is a *child* of the Master *module*.

The Rep associated with each *module* is its *representative* (lines 18 and 52). When the *representative* is empty, what we have is an *abstract module*. Filling in of the *representatives* of Master and Worker *abstract modules* with application-specific code results in the corresponding *modules*. In other words, an *abstract module* is a *module* without application-specific code.

The Master *module* interacts with each replicated Worker using its *internal communication/synchronization protocol*, PROT_Repl. The primitive operations SendWork(...), ReceiveWork(...), ReceiveResult(...),etc., (lines 27 and 54) are member functions of PROT_Repl. Each Worker *module* interacts with its *parent*, i.e., the Master *module*, using its *external communication and synchronization protocol*, which is PROT_Repl in this case (line 54).

Frame (lines 21 and 3) is a user-defined object whose data attributes can be marshaled, shipped over a communication link, and then un-marshaled, without the usual hassles of data packing and un-packing.

The example illustrates the use of a textual specification language. The specification language helps the user bypass certain C++-related and other monotonous details. As it will be evident in the next section, a user can directly write his code in C++, if desired.

```
1:  GLOBAL{
2:  #include "ImageDef.h"
3:     class Image: public UType
4:     {
5:        // A class definition specifying the attributes of an image.
6:        ...
7:     };
8:     // Similarly, global definitions of ProcessImage(...) and
9:     // WriteImage(...) go here. These procedures are used in the
10:    // following code segments.
```

```
11: }
12:
13: // The "Master" module for dynamic-replication  based computation.
14: Master EXTENDS ReplicationSkeleton
15: {
16:    // The dynamically replicated children of "Master"
17:    CHILDREN = Worker;
18:    Rep {
19:         int Number_of_Images = 0;
20:         int success;
21:         Image Frame;
22:         while (True){
23:              success = True;
24:              while ((Number_of_Images < MAXIMAGES) && success){
25:                   Produce(Frame);
26:                   Number_of_Images++;
27:                   success = SendWork(Frame); // Keep sending work-loads
28:              // to workers until none is free and can no  longer spawn  one
29:              // dynamically. SendWork is a member function of  the internal
30:              // protocol: PROT_Repl.
31:                   }
32:              if (!success) { // Do it  myself, if failed to assign work
33:                             // to a worker.
34:                   ProcessImage(Frame);
35:                   WriteImage(Frame);
36:              }
37:              if (Number_of_Images == MAXIMAGES) break;
38:         }
39:    }
40:    LOCAL {
41:       // Local functions used by this module go here.
42:       void Produce (Image& Frame)
43:       {
44:          // User code for "Produce"  goes here:
45:       }
46:    }
47: }
48:
49: // Each replicated "Worker" module.
50: Worker EXTENDS SingletonSkeleton
51: {
52:    Rep {
53:         Image Frame;
54:         ReceiveWork(Frame); // ReceiveWork is a member function of
55:                             // the external protocol: PROT_Repl.
56:         ProcessImage(Frame);
57:         WriteImage(Frame);
58:    }
59: }
```

2.2   The Model

An *architectural skeleton* is a collection of attributes which encapsulate the structure and the behavior of a parallel design pattern in an application independent manner. User extends a skeleton by filling in the structural and behavioral parameters as needed by the application at hand. Figure 1(a) roughly illustrates the various phases of application development using architectural skeletons. As shown in the figure, different extensions of the same skeleton can result in somewhat different *abstract parallel computing modules* (abbreviated as an *abstract module*). An abstract module is yet to be filled in with application code. Once an abstract module is supplied with application code, it results in a *concrete parallel computing module* (abbreviated as a *concrete module* or simply a *module*). A parallel application is a systematic collection of mutually interacting, instantiated modules.

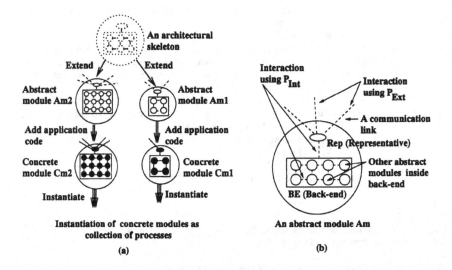

Fig. 1. (a) Relationships between an architectural skeleton, an abstract module and a module. (b) Structure of an abstract module

An abstract module inherits all the properties associated with a skeleton. Besides, it has additional parameters that depend on the needs of a given application. In object-oriented terminology, an architectural skeleton can be described as the *generalization* of the structural and behavioral properties associated with a particular parallel design pattern. An abstract module is an application-specific *specialization* of a skeleton.

Irrespective of the pattern type, an abstract module, A_m, inherits (from an architectural skeleton) the following generic set of attributes. Figure 1(b) diagrammatically illustrates the attributes of an abstract module that extends a data-parallel architectural skeleton designed for 2-D mesh topology.

- *Rep* is the representative of the abstract module. When filled in with application code, *Rep* represents the module in its action and interaction with other modules.

- *BE* is the back-end of the abstract module. Formally, $BE = \{A_{m1}, A_{m2},...,$ $A_{mn}\}$, where each A_{mi} is itself an abstract module. Note that the notion of a collection of abstract modules inside another abstract module results in a tree-structured hierarchy. Consequently, each A_{mi} is called a *child* of A_m, and A_m is called the *parent*. Abstract modules belonging to the same back-end are *peers* of one another.

- *Topology* is the interconnection topology specification of the abstract modules inside the back-end, and their connectivity specification with *Rep*.

- P_{Int} is the internal communication/synchronization protocol of the abstract module. The internal protocol is an inherent property of the associated skeleton and it captures the parallel computing model of the corresponding pattern as well as the topology. Using the primitives inside P_{Int}, the representative of A_m can interact with the abstract modules in its back-end, and an abstract module in the back-end can interact with its peers.

- P_{Ext} is the external communication/synchronization protocol of the abstract module. Using the primitives inside P_{Ext}, an abstract module (i.e., its representative) can interact with its parent and the peers. Unlike P_{Int}, which is an inherent property of the skeleton, P_{Ext} is adaptable, i.e., the module A_m adapts to the context of its parent by using the internal protocol of its parent as its external protocol. For example, in our previous master-worker example, PROT_Repl is used as P_{Ext} for the Worker module and as P_{Int} for the Master module.

Though an abstract module is an application specific specialization of an architectural skeleton, it is still devoid of any application code. User writes application code for an abstract module using its communication/synchronization protocols, P_{Int} and P_{Ext}. A code-complete abstract module is called a *concrete parallel computing module*, or simply a *module*. A parallel application is a hierarchical collection of instances of the modules.

As mentioned before, the notion of parent-child relationships among modules result in a tree-structured hierarchy. A parallel application can be viewed as a hierarchical collection of modules, constituting of a *root* module and its children forming the sub-trees. This tree is called the *hierarchical tree* associated with the application. For instance: (1) in the previous example, the Master module forms the root of the hierarchy, and the dynamically replicated children Worker modules form the sub-trees. (2) In an application consisting of the three modules: Producer, Worker and Consumer, a *compositional module* forms the root of the hierarchy, and its three children (i.e., Producer, Worker and Consumer) form the sub-trees.

The concept of the hierarchical tree is important, because the OO implementation dynamically constructs the hierarchy associated with an application, while completely hiding it from the user. A *singleton module*, which has no children, forms a leaf in the hierarchy.

3   An Object-Oriented Implementation

The model is presently implemented in industry standard C++ (SunCC compiler, V 4.1), without requiring any language extensions. MPI is used as the underlying communication library. Work is continuing for porting it to GNU C++, and the C++ compiler for AIX (i.e., the xlC compiler) on RS6000.

A specification-language based textual interface, whose parser is implemented in PERL, helps the user in the various stages of application development. It is parsed to produce the back-end C++ code. However it should be emphasized that use of a specification language is not a language extension. It merely helps the user to bypass certain monotonous C++-based details that can easily be generated using PERL. An expert in C++, for example, may want to directly develop his application in C++.

Other important features of the implementation include: (1) use of C++ operator-overloading to implement certain primitives inside protocol classes (e.g. Send(...) and Receive(...) inside PROT_Net); (2) use of marshaling and un-marshaling mechanisms whereby the data attributes of an entire object can be marshaled, shipped over a communication link and then un-marshaled, without the usual hassles of data packing and unpacking.

3.1   The Specification Language Parser

The example in the previous section illustrates the current user's interface which uses a specification language. The specification language parser, implemented in PERL, translates it to produce the back-end C++ file: *Pmain.cc*. It is subsequently compiled and linked with the skeleton library to produce the executable file, designated to run on a workstation cluster.

The following is the skeleton of the automatically generated file *Pmain.cc*. As mentioned before, a user can directly write his code in C++. The specification language and its parser merely reduce some of the extra work by automating certain monotonous steps.

```
#include "BasicDef.h"
#include "SingletonSkeleton.h"
// Similarly other "include" files go here.
//Any global definitions will go below:
//---------------------------------------
#include "ImageDef.h"
  class Image: public UType
  {
    // A class definition specifying  the attributes of an image.
  };
  // Similarly, the other global definitions inside "GLOBAL" follow:
//---------------------------------------------

// Generated code for module: "Worker"
class Worker : public SingletonSkeleton <PROT_Repl>
{
```

```
public:
    Worker(Void& _v) {};
    virtual void Rep() { // The representative code goes here. }
    // Miscellaneous local definitions go below:
};
// Generated code for module: "Master"
class Master : public ReplicationSkeleton <Worker, PROT_Repl, Void>
{
    Master() {};
    virtual void Rep() { // The representative code goes here.  }
    // Miscellaneous local definitions go below:
};
void Pmain()
{
    Master TopLevel_366;
    TopLevel_366.Run();
}
```

As is evident from the previous code, C++ templates are used extensively in the implementation. PROT_Repl is the internal protocol for the replication skeleton. Consequently, PROT_Repl becomes the external protocol for a Worker module. Since the Master is at the root of the hierarchy, its external protocol is undefined (as specified by Void).

3.2   Implementing Architectural Skeletons

Figure 2 illustrates the high-level class diagram behind the design of the skeleton library, using the standard UML notation. For simplicity, the figure does not illustrate the relationships between the skeleton classes and the various protocols. Besides, the various attributes and the methods associated with each class, and the formal parameters associated with each inherited skeleton class are not shown for a cleaner representation.

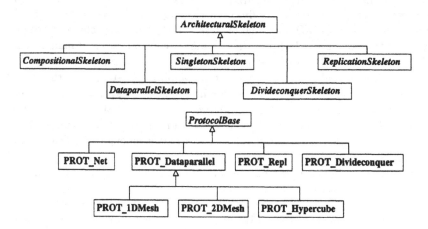

Fig. 2. High level class diagram for the skeleton-library

Figure 3 shows more details of the high level design pertaining to the replication and the singleton skeletons, using UML notation. These skeletons are used in our master-worker example in Sect. 2.1. The figure also illustrates the high level class diagram of these skeletons as used in the example. As it is evident from the figure, each skeleton class is implemented as a template, with its list of formal parameters. A template class is subsequently parameterized with actual application-specific value-list to create a bound-class, which is further extended by the user.

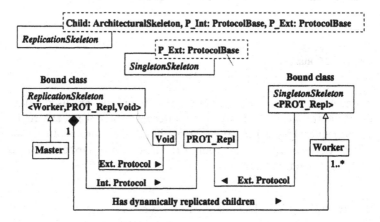

Fig. 3. High level class diagram for the example

3.3   The Dynamic Execution Model

After viewing the static class diagram of the skeleton library and the application, we are in a position to briefly discuss the dynamic execution model. The execution model is SPMD, i.e., each processor in the processor-cluster loads and executes the same file, which results in major savings in terms of management of source, object and executable files. Consequently, each process falls through the same hierarchical tree associated with the application, starting at the root of the tree. Figure 4(a) illustrates the hierarchical tree associated with the master-worker example discussed in Sect. 2.1. Figure 4(b) illustrates the hierarchical tree associated with a hypothetical Producer-Worker-Consumer application. Here, the Worker module further sub-divides its work-loads among dynamically replicated sub-workers.

A node of a hierarchical tree is essentially the representative of a module. Each process is responsible for executing exactly one node of the tree, i.e., there is a one-to-one correspondence between a process and a representative node. Each process starts at the root of the hierarchy and then traverses down the tree to its designated node.

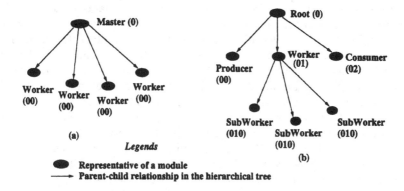

Fig. 4. Hierarchical tree and its traversal scheme

How does a process know, which path in the tree to traverse? This is achieved as follows: each process is dynamically assigned an identification string, by following a unique labeling scheme. As a process traverses down the tree, it dynamically calculates its path, by following the same scheme. The process traverses down a specific path of the tree, if and only if the already calculated path is a substring of its assigned identification string. When the calculated path matches the identification string, the process is at its designated node.

In Fig. 4, the printed string inside parentheses pairs beside each node is precisely the identification string of the process which executes it. The dynamically replicated Workers in the first application are all identical, therefore they execute the same code and have the same identification string. The same situation applies to the dynamically replicated SubWorkers in the second application.

All of the above issues are completely hidden from the user. In fact, for instance, a user follows the general structure as illustrated by the master-worker application in Sect. 2.1, and writes his application with the perspective that he is dealing with individual parallel computing modules, rather than with individual processes. Without any further aid from the user, the dynamic execution model makes it possible for a process to execute the code segment pertaining to a given parallel computing module.

4 Performance Results

Experiments were conducted to assess the performance of the system. The results were compared with direct MPI-based implementations. The performance difference is within ±5%, which can be attributed to the fact that the skeleton-library is implemented as an extremely thin layer on top of MPI. The results obtained for the PQSRS algorithm [8] and the 2-D discrete convolution algorithm [9] are described in this section. Detailed results with different applications can be found in [10].

PQSRS (Parallel Quick Sort using Regular Sampling) is a parallel version of quick sort, shown to be effective for a wide variety of MIMD architectures. It uses a master-slave pattern, which is easily realized using the data-parallel skeleton for mesh topology and the singleton skeleton (analogous to the structure of the example in Sect. 2.1). The algorithm works in the following steps: (1) the master partitions the data items to be sorted to the N children (i.e., slaves). Each child then performs sequential quick sort on its own data items, selects N data items as regular samples, and sends them back to the parent (i.e., master). (2) the master gathers the regular samples from all its children, sorts them, gathers $N - 1$ pivot values and broadcasts them to the children. Each child partitions its portion of sorted items into N disjoint partitions, based on the $N - 1$ pivot values. (3) Child i keeps the ith partition and sends the jth partition to its jth peer. Thus, at this phase, each child has to communicate with all its $N - 1$ peers. (4) Each child receives $N - 1$ partitions from its peers, merges them with its own partition to form a single sorted list, and sends the sorted list back to the master. Finally, the master concatenates the sorted sub-lists from all its children to form the final sorted list.

PQSRS is a non-trivial algorithm which requires a considerable amount of peer-to-peer interaction among the slaves, which is supported by the internal protocol(s) of the data-parallel skeleton. It cannot be implemented using most other pattern based systems. The second set of experiments involves a relatively trivial application, called the 2-D discrete convolution algorithm. This is an image processing algorithm, where a mask is applied to the image pixels to produce a convoluted image. Like most other image processing algorithms, it follows the master-slave pattern and can be elegantly implemented using the same set of skeletons as in the example presented in Sect. 2.1. Unlike PQSRS, no peer-to-peer interaction among slaves is needed for this specific application.

Figure 5 illustrates the results obtained for sorting 10000 randomly generated objects using PQSRS, and for the discrete convolution of a 400×400 pixel image using a 5×5 mask. The underlying hardware is a cluster of Sun Sparc workstations connected by a 10-megabit ethernet network. For the discrete convolution algorithm, the speed-up ratio was measured with respect to the best sequential algorithm. In the case of PQSRS, the speed-up ratio was measured with respect to the same sequential quick-sort routine used inside PQSRS. As it turns out in both the applications, the granularity becomes too small with more than 10 processors and hence the performance gradually degrades.

5   Conclusion

The paper presents a generic model for designing and developing parallel applications, which is based on the idea of design patterns. Patterns are abstracted as architectural skeletons. The model is an ideal candidate for implementation using object-oriented techniques. The object-oriented approach can be used to build application-independent library of skeletons, while keeping in mind flexibility and extensibility as two of the major issues. The present set of architectural-skeletons supports patterns for coarse-grain message-passing computation that

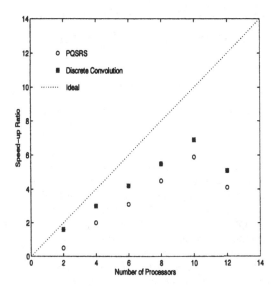

Fig. 5. Speed-up ratio versus number of processors

provide good performance in a networked MIMD environment. Incorporation of new skeletons for such an environment is an ongoing research activity.

References

1. Gamma, E., Helm, R., Johnson, R., Vlissides, J.: Design Patterns: Elements of Reusable Object-Oriented Software. Addison-Wesley Publishing Company (1994)
2. Singh, A., Schaeffer, J., Szafron, D.: Experience with parallel programming using code templates. Concurrency: Practice and Experience. (1998) 10(2):91–120
3. Browne, J.C., Hyder, S.I., Dongarra, J., Moore, K., Newton, P.: Visual Programming and Debugging for Parallel Computing. IEEE Parallel and Distributed Technology. (1995) 3(1):75–83
4. Siu, S., Singh, A.: Design Patterns for Parallel Computing Using a Network of Processors. In Sixth IEEE International Symposium on High Performance Distributed Computing, Oregon, USA. (1997) 293–304
5. Cole, M.: Algorithmic Skeletons: Structured Management of Parallel Computation. The MIT Press, Cambridge, Massachusetts (1989)
6. Campbell, D.K.G.: Towards the Classification of Algorithmic Skeletons. Technical Report YCS 276, Department of Computer Science, University of York (1996)
7. Pandey, R., Browne, J.C.: A Compositional Approach to Concurrent Programming. In Proc. of 1996 International Conference on Parallel and Distributed Processing Techniques and Applications (PDPTA'96), California. (1996) 1489–1500
8. Quinn, M.J.: Parallel Computing: Theory and Practice. McGraw-Hill, Inc. (1994)
9. Myler, H.R., Weeks, A.R.: The Pocket handbook of Image Processing Algorithms in C. Prentice Hall (1993)
10. Goswami, D.: A Design Pattern Based Approach for Developing Parallel Applications. PhD thesis, Department of Electrical and Computer Engineering, University of Waterloo. (In preparation)

Loci: A Deductive Framework for Graph-Based Algorithms*

Edward A. Luke

Mississippi State University, Mississippi State, MS, USA

Abstract. Modern computational fluid dynamics (CFD) software is complex. Often CFD simulations require complex geometries, flexible boundary conditions, multiple integrated computational models (for example, heat conduction, structural deformations, gas dynamics, etc.), as well as grid adaptation. As a result of this complexity, the correct implementation of numerical simulation components is actually less challenging than guaranteeing the correct coordination of complex component interactions. If one is to consider the development of CFD applications that reliably incorporate a broad selection of numerical models, then one must consider technologies that simplify, automate, and validate the numerical model coordination mechanisms. The Loci system presented here addresses these issues by introducing a deductive framework for the coordination of numerical value classes constructed in C++.

1 Introduction

Recent developments in object-oriented numerics can roughly be classified into two broad branches: 1) the development of flexible high performance value classes such as MTL[1], Blitz++[2], and POOMA II [3], and 2) the development of application level frameworks or toolkits such as PetSc [4], and LPARX [5]. High performance value classes have significant reuse potential since their abstractions are generally built from well known and often used mathematical constructions such as vectors, matrices, tensors, and so on. On the other hand, application toolkits tend to be more specialized (for example, LPARX attacks AMR algorithms, while PetSc tends to be Krylov subspace centric). These two approaches are addressing fundamentally different problems: value classes reduce complexity associated with representing the mathematical structure (equations) of application components while application toolkits reduce complexity of integrating a diverse set of loosely related components.

The Loci[1] system is an application framework that seeks to reduce the complexity of assembling large-scale finite-difference or finite-element applications, although it could be applied to many algorithms that are described with respect to a connectivity network or graph. The design of the Loci system recognizes

---

* This work was partially supported by NSF Cooperative agreement number ECD-8907070
[1] The name Loci is derived from the framework's ability to compute the locus of application component specifications.

S. Matsuoka et al. (Eds.): ISCOPE'99, LNCS 1732, pp. 142–153, 1999.

that a significant portion of the complexity and bugs associated with developing large scale computational field simulations is derived from errors in control and data movement. In other words, a significant number of errors in these applications are caused by incorrect looping structures, improper calling sequences, or incorrect data transfers. Many of these problems are subtle and result from gradual evolution of the application over time giving rise to inconsistencies between various application components. The Loci framework addresses these problems by automatically generating the control and data movement operations of an application from component specifications while guaranteeing a level of consistency between components. The approach taken is similar to Strand [6] except that it includes semantics of unstructured mesh computations in rule specifications.

2   The Loci Data Model

In the Loci system, computational graphs are represented by collections of entities and collections of maps or connectivity lists. Entities represent sites where computations may occur in the graph. For example, the entities in a finite-volume calculation may represent faces, cells, and nodes of the mesh. Maps may connect faces to their left and right cells, or cells to their nodes, and so forth. Values are bound to entities via the **store** construct which provides an injective mapping from entities to values. The **parameter** construct provides a singleton interface to value where a set of entities is mapped to a single value. Relationships between entities are provided by the **map** construct. The map construct can be composed with the store construct to provide an abstraction of indirection. The **constraint** construct, used to constrain computations to some subset of entities, provides an identity mapping over a given subset of entities. These constructs are illustrated in figure 1.

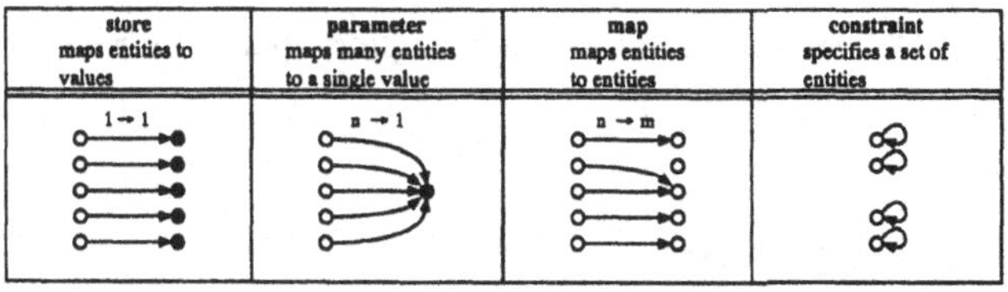

store maps entities to values	parameter maps many entities to a single value	map maps entities to entities	constraint specifies a set of entities

Fig. 1. Four basic database constructs

These basic constructs are used to formulate a database of facts that describe the problem. Each fact provides information about some subset of entities, such as positions of nodes, or maps relating cells to nodes. Each of these facts is given an identifier that consists of a name, an iteration label, and an iteration offset. The iteration label corresponds to the nested iteration levels of a loop as in figure 2. Iteration labels form a partial order that is grounded at the stationary

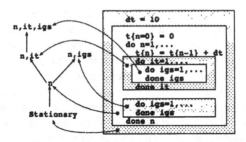

Fig. 2. Nested iterations as a hierarchy

(constant) iteration level. Iteration offsets provide a means of accessing information at a previous iteration. The general form of a fact identifier is $\alpha\{\tau + \theta\}$ where α is the name, τ is the iteration identifier and θ is the iteration offset. For example, `pressure{n-1}` represents the identifier for pressure at the previous iteration of iterator n.

The "->" operator is used to represent the application of a map in the access of values. Thus the indirection typically encountered in unstructured calculations can be encoded: accessing the value for the left side of a face identified by entity f may be coded using C arrays as `value[left[f]]`, whereas in Loci this access of value through the mapping `left` is written as `left->value`. Figure 3 illustrates the aggregate perspective of the indirection operation represented by the "->" operator.

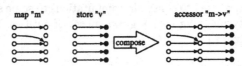

Fig. 3. An illustration of the mapping operation 'm->v'

3   Rule Specifications

In addition to a database of facts that includes the problem specification, a database of rules describes transformations that can be used to introduce new facts into the database. These rules correspond to fundamental computations involved in solution algorithms such as rules for evaluating areas of faces, or for solving equations of state. These rules are specified using text strings called rule signatures that describe the input stores, parameters, and maps required to perform a computation and the list of stores or parameters that it generates. Rule signatures are of the form head <- body where head consists of a list of variables that are generated by the application of the rule, while the body contains a list of variables that are accessed while performing the computation. For example, the rule signature `p<-rho, T, R` represents that a value for pressure (p) is provided when values for density (rho), temperature (T), and gas constant (R) are present.

A rule signature may also contain the mapping operator "->" to represent the use of indirection in a computation. For example, the rule signature of a computation that generates areas of faces utilizing positions of related nodes may be given as `area<-face2nodes->position`, where "face2nodes" is a mapping that connects faces to their defining nodes. Note that this rule may represent a calculation for a single face or for all of the faces in the mesh. It states that for all entities that satisfy all the properties given in the body, a computation can be performed to produce a value given in the head.

There are several classes of computations that are encountered in a typical unstructured grid computation. This section will identify the semantics of these computation rule classes.

3.1 Rule Constraints

Any rule that is specified can be constrained to only compute values for some subset of entities. For example, if one wanted to specify the value of variable u at nodes where the Dirichlet boundary condition are applied, this might be specified by the rule signature: `u<-CONSTRAINT(Dirichlet)`. In addition to constraining rule applications, constraints also provide assertion semantics: a constraint implies that a rule must provide values for every entity in the constraint. In many cases this can be used to automatically detect inconsistencies caused by incomplete specifications. In addition, Loci identifies over-specification by requiring that each entity may have only one given value. Thus, if a boundary condition is applied to an interior node of the domain, then an error would result caused by a conflict between the boundary condition and stencil specifications.

3.2 Point-wise Rules

The most common rule in finite-difference or finite-element applications is the point-wise rule. The point-wise rule represents an entity by entity computation of values that are placed in the stores listed in its head. The computation specified in the rule is referentially transparent and local. The semantics of the point-wise rule application requires that an output variable can only define one value per entity. This is a variation of the single assignment semantic. If an ambiguous specification produces two rules that compute values for the same entity, it is flagged as an error during scheduling. Recursion is allowed in point-wise rules, provided that the single value per entity rule is not violated. Thus, recursion in point-wise rules is bounded to the number of entities in the simulation mesh. An example of a symmetric Gauss-Seidel method implemented through point-wise rule recursion will be illustrated in Sect. 6.2.

3.3 Reduction Rules

The Bird-Meertens formalism[7] provides a means of describing what are otherwise known as reduction operations[2]. In this abstraction, a reduction is described by a function composed of three components: a function that is applied to a set of values, an associative and commutative operator $\oplus$ that is defined on the type returned by the above mentioned function, and an identity element of operator $\oplus$, e. Thus a reduction, r, over values, $\{v_i | i \in [1, N]\}$, using function f and operator $\oplus$ is defined as

$$r = f(v_1) \oplus f(v_2) \oplus \cdots \oplus f(v_i) \oplus \cdots \oplus f(v_N). \tag{1}$$

When the reduction is evaluated using a left or right precedence rule, then a sequential evaluation is derived; however, the associative property of $\oplus$ allows for different parallel evaluation orders. For example, the set of values can be partitioned into subsets that can be evaluated concurrently as in

$$r = \{e \oplus f(v_1) \oplus f(v_2) \cdots \oplus f(v_p)\} \oplus \{e \oplus f(v_{p+1}) \oplus \cdots \oplus f(v_N)\}. \tag{2}$$

Note that the addition of the identity element, e, is used to indicate the requirement for the initialization of value on each processor: all reductions begin with the identity element. Parallel partitioning of reduction operations can be expressed when given three basic computational methods identified as unit, apply, and join as listed in Table 1. The unit rule initializes a reduction variable to the identity element, the apply rule "accumulates" results of a function application to a set of values, and the join operation "accumulates" the partial results. The algorithm for partitioning this reduction operation among parallel processors is accomplished by creating a copy of the reduction variable on each processor participating in the reduction. Each reduction variable is initialized to the identity, and then followed by the application of all apply rules that are in that processor's partition. Finally the partial results for each processor are reduced to the final result using join operations.

Table 1. Reduction specification in Loci

Rule Type	Function	Rule Signature
Unit Rule	$r_i^0 = e$	$r \leftarrow CONSTRAINT(v), UNIT(e)$
Apply Rule	$r_i^{j+1} = r_i^j \oplus f(v_i)$	$r \leftarrow r, v, APPLY(\oplus)$
Join Op	$r_i^{m+n} = r_i^m \oplus r_i^n$	Derived (No signature)

---

[2] These reduction operations may be computations of global simulation parameters such as the maximum stable time-step, or of local accumulations such as summing mass contributions from cells to nodes.

3.4 Iteration Rules

Iteration is defined by way of three types of rule specifications: build rules that construct the iteration, advance rules that advance the iteration, and collapse rules that terminate the iteration. This specification follows an analogy to the inductive proof in that build rules are analogous to an inductive base while advance rules are analogous to an inductive hypothesis.

For example, an iteration where a variable named q is iterated to a converged solution may be described by the following three rules: 1) a build rule of the form q{n=0}<-ic, 2) an advance rule similar to q{n+1}<-q{n},dq{n}, and 3) an iteration collapse rule solution<-q{n},CONDITION(converged{n}). Iteration in this example proceeds by initializing the first iteration, q{n=0}, using the build rule. Next, termination of iteration is then checked by computing converged, if the test succeeds then the collapse rule terminates the iteration. Finally the iteration advances in time by the repeated application of the advance rule which computes values for q for the next iteration ({n+1}) given current iteration values at time level {n}. Note that the completion of these rules may require invoking other rules specified in the rule database. In this case, rules that compute converged{n} and dq{n} will also need to be scheduled.

To support iteration, variables that exist in lower levels of the iteration hierarchy are automatically promoted up the iteration hierarchy. Thus a variable that is computed in iteration {n} is communicated to iteration {n,it} automatically. In addition, rules that are specified completely at the stationary level will be promoted to any level of the hierarchy. This allows for the specification of relations that are iteration independent (for example, $p = \rho \tilde{R} T$ implies $p^n = \rho^n \tilde{R}^n T^n$).

4 Scheduling

In Loci, the mesh, boundary conditions, initial conditions, and other modeling information is stored in a database of facts using stores, parameters, maps and constraints. The application provides a set of rules that may be used to solve the given problem. Given these databases, an application is formed by searching for a goal specified by the user. This basic strategy is illustrated in figure 4. The default goal is to compute a variable called solution, which plays a role similar to main in C++. The search for this goal may yield three possible results: 1) there is no way to obtain the request from the given database, 2) the request can be satisfied, but not without ambiguous results or unsatisfied constraints (assertion failures), or 3) the request can be satisfied and this is the resulting schedule that generates this goal. Generally, case 1 is caused by insufficient information in the given fact database (under-specification), while case 2 is caused by internal inconsistencies that usually result from conflicting database specifications (over-specification).

Scheduling occurs in three steps. The first step involves creating a dependency graph that connects the variables stored in the fact database to the goal using

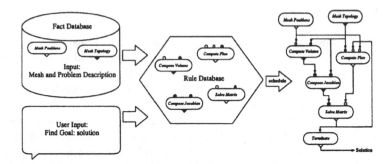

Fig. 4. The Loci methodology for automatic application generation.

the rules in the rule database. This step involves iteratively exploring the space of known variables and determining which rules can apply, which may in turn generate new variables. Once this graph is produced it is pruned to only those rules that generate the requested goal, sorted into iteration hierarchies, and reduced to a directed acyclic graph (DAG) by clumping recursive dependency loops. The next step is an existential deduction phase which determines what attributes can be assigned to which entities. For example, the rule p<-rho,R,T specifies that entities that have attributes rho, R, and T also have the attribute p. The existential deduction begins with the given facts, and follows the DAG order computing the entities associated with each attribute until the goal is reached. During this existential deduction, recursive loops are iteratively evaluated until all possible attributes are generated. The result of the existential deduction phase is a concurrent schedule that obtains the requested goal. However, since it is possible that some attributes may exist for entities that do not contribute to the requested goal, a final optimization pass prunes this schedule. The pruning operation starts from the goal and works backwards through the DAG until the schedule only computes those values that are needed to provide the requested goal.

The scheduling process naturally produces a concurrent schedule. Only partitioning of entities to processors is needed to generate a schedule for parallel processors (on distributed memory architectures a communication schedule would also need to be deduced). Thus the numerical model does not have any references to parallel execution – this arises naturally from the specification.

5   Implementation

The fundamental design strategy in the Loci system has been the use of shallow inheritance hierarchies combined with templated containers and composers. The most basic data type for the Loci system is the entitySet, a value class that describes arbitrary sets of entities, and provides fast intersection, union, and complement operations. These entity sets are necessary for the existential deduction phase of scheduling and used for control and allocation functions.

The data models described in Sect. 2 are implemented as templated container classes. These containers provide features that facilitate their storage in the fact database and their automatic binding in rule invocations. The rules described in Sect. 3 are implemented via a shallow inheritance hierarchy. Users create new rules by providing a constructor that creates the rule specification and a virtual member function called compute that performs the specified computation.

A high-performance implementation strategy is based on moving all run-time polymorphic behavior (virtual functions and RTTI) to the level of collections of entities. This is facilitated by the scheduler which generates a sequence of rule applications over sets of entities. Each rule implementation provides a virtual member function compute that provides a collection level interface to rule evaluation. The scheduler generates a partial order of entity computations for each rule invocation represented by the sequence class. The developer of a rule implementation provides an inlined member function that performs the specific computations required by a single entity. A templated composer function, do_loop, provides the interface between the per-entity computation provided by the user and the collection-level request made by the execution schedule (in the form of an instance of sequence). This strategy provides a powerful division of labor with respect to optimization: while the scheduler can concern itself with global analysis such as the identification of concurrent or cache optimized evaluation orderings, the do_loop composer can provide the optimal looping interface (e.g. loop unrolling or parallelization directives). Most importantly, all of these optimizations can occur without change to the numerical algorithm as specified in the rule definition.

6 Examples

The Loci system is currently being used to develop a finite-rate chemically reacting compressible flow solver for simulation of high speed and combustion flow problems. The code presently uses an implicit finite-volume scheme for three dimensional unstructured mixed element grids using flux difference splitting schemes for stable solutions of flow fields containing discontinuities such as shocks. This application is currently being used at the NASA Stennis space center to compute hydrogen oxygen combustion reactions to assess rocket motor measurement strategies. Two examples are provided from this flow solver implementation to illustrate some of Loci's features.

6.1 Area Computation

A generalized unstructured grid may consist of hexahedra, tetrahedra, prisms, and pyramids, which are in turn composed of triangular and quadrilateral faces. The finite-volume integration method requires surface integrals of these elements which in turn requires an area evaluation of these faces. Two separate rules are required to generate areas for the quadrilateral and triangular faces. The quadrilateral faces are defined by the map variable qnds, while the triangular faces are defined by the map variable **tnds**. Nodal position vectors are stored in **pos**. Figure 5 illustrates the implementation of a Loci rule that computes the area of a quadrilateral face. In this rule, the area is determined by the cross product of the two diagonal vectors of the face. The constructor for this rule creates the specification "**area<-qnds->pos**", while method **compute** provides an interface for the actual computation of areas.

```
class quad_area : public pointwise_rule {      inline void quad_area::calculate(Entity fc) {
  const_store<vect3d> pos ;                       vect3d dv1 = pos[qnds[fc][2]]-pos[qnds[fc][0]];
  const_MapVec<4> qnds ;                          vect3d dv2 = pos[qnds[fc][1]]-pos[qnds[fc][3]];
  store<Area> area ;
public:                                           area[fc].n = cross(dv1,dv2) ;
  quad_area() ;                                   real sada = sqrt(dot(area[fc].n,area[fc].n)) ;
  void calculate(Entity fc) ;                     area[fc].n *= 1./(sada+EPSILON) ;
  virtual void                                    area[fc].magnitude = 0.5*sada ;
      compute(const sequence &seq) ;            }
} ;

quad_area::quad_area() {                         // calculate areas for sequence of entities
  name_store("pos",pos) ;                        void quad_area::compute(const sequence &seq) {
  name_store("qnds",qnds) ;                        do_loop(seq,this,&quad_area::calculate) ;
  name_store("area",area) ;                      }

  input("qnds->pos") ;                           // register rule in global rule database
  output("area") ;                               register_rule<quad_area> quad_area_registration ;
}
```

Fig. 5. Quadrilateral-face area computation rule in Loci

6.2   Gauss-Seidel Iteration

The linear system solution required by the implicit time integration in the solver utilizes a Gauss-Seidel iterative method. This method solves the equation $Ax = b$ by iteratively evaluating the equation $x^{k+1} = (L+D)^{-1}(Ux^k + b)$, where $A = L + D + U$, and L, D, and U are the lower, diagonal and upper matrices respectively. This method can be written as the point-wise recurrence relation

$$x_i^{k+1} = D_i^{-1} \left(b_i - \sum_{j=1}^{i-1} L_{ij} x_j^{k+1} - \sum_{j=i+1}^{n} U_{ij} x_j^k \right) .$$

In the unstructured flow solver, the matrix, A, is formed from the connections between cells. These connections represent faces in the mesh. The data structure is constructed by first identifying a left and right cell to every internal face with maps **cl** and **cr**. The faces are constructed such that **cl** always points to the lowest color cell (in the matrix coloring). Mappings to the lower and upper parts of the matrix are formed by inverting the maps **cl** and **cr** such that lower = $inverse(\text{cr})$ and upper = $inverse(\text{cl})$.[3] In this data structure, the components of the lower matrix are stored at faces in variables called L and U while the diagonal is stored at cells in a variable called D. A more detailed description of

---

[3] An inverse of a mapping is simply its transpose.

the matrix data-structure is provided in [8]. Using this data structure, the above recurrence relation form of the Gauss-Seidel algorithm becomes the recursive point-wise rule

`x{k+1}<-D,b,lower->L,lower->cl->x{k+1},upper->U,upper->cr->x{k}.`

Loci schedules recursive rules by repeatedly evaluating the rule until all possible entities are computed. What does Loci do for a red-black coloring? In this case all red cells have only left faces pointing to them, while all black cells have only right faces pointing to them as illustrated in figure 6. Thus red cells have a null mapping for lower, while black cells have a null mapping for upper. In the first step of the repeated evaluation, Loci will be able to evaluate this rule only for red cells since the lower map is null only for these cells and no entities have the attribute x{k+1}. For the second iteration of this scheduling process Loci will be able to apply this rule for all black cells since the red cells produced the necessary x{k+1} attribute. A following iteration will determine that no new attributes can be assigned and the schedule will be complete. A two step concurrent red-black algorithm will be the

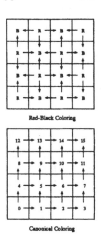

Fig. 6. Mesh colorings

result. Interestingly, if instead a regular mesh is given with a canonical coloring, then Loci will derive the concurrent hyper-plane algorithm used in the NAS parallel benchmark LU[9].

7   Performance

Evaluating the performance of the finite-rate chemistry application developed in Loci is complicated by the fact that there is no application which exactly matches its particular design goals. It is important, however, to obtain a performance baseline to determine if the approach has an unacceptable performance penalty. Unfortunately, simple linear algebra based benchmarks over-predict performance due to their unrealistically high percentage of multiply-add operations. In contrast, the finite-rate chemistry solver spends less than 20 percent of its execution time in matrix solution. To provide a more reasonable performance baseline, the performance of a Fortran ideal-gas solver[10] is measured. The basic algorithms used in both solvers are similar, although a more robust, but more expensive, method for Jacobian computations is applied in the finite-rate chemistry application.

The performance of a serial Loci implementation is measured for the finite-rate chemistry application using the hardware counters on the SGI 195mhz R10K processor. A medium sized grid for a converging diverging rocket nozzle consisting of 15,488 cells is used in the performance measurement. Two chemistry models are used in the simulation: 1) an ideal gas model, and 2) a 6 species

28 reaction hydrogen-oxygen reaction kinetics model. The performance of the Fortran ideal-gas solver is also measured to provide a performance baseline. The performance is based on the ratio of floating-point operations obtained from the R10K hardware counters to an overall execution time including Loci schedule generation. Since the operations performed by scheduling is not included in the operation measurement, scheduling cost can only decrease measured performance. These results, listed in Table 2, indicate that a deductive scheduling system combined with C++ templates and value classes can produce performance similar to Fortran based applications. The lower performance of the ideal-gas model is a result of a design that supports flexible chemistry models rather than particular overheads of the Loci system.

Table 2. Loci performance results on nozzle simulation

Code	Model	Flops	Time (sec)	Rate (M-Flops)
Loci	Ideal Gas	41.41e9	1174	35.3
Loci	H2-O2 Kinetics	252.89e9	5803	43.6
Fortran	Ideal Gas	33.80e9	749	45.1

8 Summary

This paper outlined a programming model based on a data parallel logic programming strategy. The programming model has the advantage that it can coordinate various numerical models while enforcing many consistency guarantees. A side effect of this model is that parallel schedules can be generated from the specification. A sequential implementation of this model in C++ demonstrates that the approach can be competitive with Fortran in performance while simultaneously providing consistency guarantees traditional design approaches lack.

The current implementation provides an application framework where high-performance value classes are used to describe entity-based computations, while the Loci scheduling system provides an automatic assembly of these fundamental descriptions into an application-level schedule. In this sense, the Loci system is both a library and a run-time system for the development of high performance graph-based applications. Although the present implementation focuses on the composition of entity-based computations, the same strategy could be applied to the case of aggregate computations (such as might be performed by a PetSc linear system solver, or an external commercial simulation application). However, these cases would need special treatment by the scheduler to ensure that the computations remained aggregate (for example, a matrix can not be inverted in segments), and may not be subject to the same parallelization opportunities.

There is a design argument that has been subtly referred to throughout this paper: in scientific computing, the needs of application-level structures are fundamentally different than that of the value-level. While recent advances in value class design are providing clear benefits, the problems associated with application-level structures have been less clearly addressed. A likely source of difficulties at this level is due to complex interactions between application components that do not fit neatly into a hierarchical, object-oriented structure, but rather appear to be more readily represented by axiomatic statements (*e.g.* pressure exists whenever density, temperature, and gas constant exist). By this argument, a multi-paradigm approach offers the best of both worlds. In the case of Loci, C++ provides powerful mechanisms for creating high performance value classes, while Loci provides a powerful logic-based scheduling infrastructure for coordinating application-level program structures.

References

1. J. G. Siek and A. Lumsdaine. The Matrix Template Library: A Generic Programming Approach to High Performance Numerical Linear Algebra. In *Proceedings of the 2nd International Scientific Computing in Object-Oriented Parallel Environments (ISCOPE'98)*, Lecture Notes in Computer Science. Springer-Verlag, 1998.
2. T. L. Veldhuizen. Arrays in Blitz++. In *Proceedings of the 2nd International Scientific Computing in Object-Oriented Parallel Environments (ISCOPE'98)*, Lecture Notes in Computer Science. ISCOPE, Springer-Verlag, 1998.
3. S. Karmesin, J. Crotinger, J. Cummings, S. Haney, W. Humphrey, J. Reynders, S. Smith, and T. Williams. Array design and expression evaluation in POOMA II. In *Proceedings of the 2nd International Scientific Computing in Object-Oriented Parallel Environments (ISCOPE'98)*, Lecture Notes in Computer Science. ISCOPE, Springer-Verlag, 1998.
4. S. Balay, W. D. Gropp, L. C. McInnes, and B. F. Smith. Efficient management of parallelism in object-oriented numerical software libraries. In E. Arge, A. M. Bruaset, and H. P. Langtangen, editors, *Modern Software Tools in Scientific Computing*, pages 163–202. Birkhauser Press, 1997.
5. Scott R. Kohn. *A Parallel Software Infrastructure for Dynamic Block-Irregular Scientific Calculations*. PhD thesis, University of California, San Diego, 1995.
6. I. Foster and S. Taylor. *Strand: New Concepts in Parallel Programming*. Prentice-Hall, 1990.
7. R. S. Bird and L. Meertens. Two exercises found in a book on algorithmics. In L. G. L. T. Meertens, editor, *Program Specification and Transformation*, pages pp 451–457. North-Holland, 1987.
8. E. A. Luke. *A Rule-Based Specification System for Computational Fluid Dynamics*. PhD thesis, Mississippi State University, December 1999.
9. E. Barszcz, R. Fatoohi, V. Venkatakrishnan, and S. Weeratunga. Solution of regular, sparse triangular linear systems on vector and distributed-memory multicomputers. Technical report, NASA Ames Research Center, Report RNR-93-007, NASA Ames Research Center, Moffett Field CA 94035, April 1993.
10. J.M. Janus. *Advanced 3-D Algorithm for Turbomachinery*. PhD thesis, Mississippi State University, May 1989.

The OptSolve++ Software Components for Nonlinear Optimization and Root-Finding

David L. Bruhwiler, Svetlana G. Shasharina, and John R. Cary

Tech-X Corporation, Boulder CO 80303, USA

Abstract. OptSolve++ is a set of C++ class libraries for nonlinear optimization and root-finding. The primary components include TxOptSlv (optimizer and solver classes), TxFunc (functor classes used to wrap user-defined functions), TxLin (linear algebra), and a library of test functions. These cross-platform components use exception handling and encapsulate diagnostic output for the calling application. Use of the "template" design pattern in TxOptSlv provides a convenient interface that allows for interchange of existing algorithms by users and straightforward addition of new algorithms by developers. All classes are templated, so that optimization over the various floating-point types uses the same source code, and so that optimization over integers or more esoteric types can be readily accommodated with an identical interface. TxOptSlv and TxFunc use the template-based "traits" mechanism, allowing them to work with any user-specified container class, and future versions of OptSolve++ will allow users to interchange TxLin with other linear algebra libraries.

1 Introduction

OptSolve++ is a set of object-oriented C++ class libraries for nonlinear optimization and root-finding. Each library is designed as a software component that can be embedded in other software and extended via inheritance to add new capabilities. The primary components include TxOptSlv (the optimizer and solver classes), TxFunc (functor classes used to wrap user-defined functions), TxLin (linear algebra classes), and a library of standard test functions. These components compile on PC, Macintosh, GNU/Linux and all major Unix platforms. Use of exception handling and the encapsulation of all diagnostic output makes them robust and reliable.

2 TxOptSlv Library of Nonlinear Optimizers and Solvers

The TxOptSlv library provides a convenient and flexible interface for nonlinear optimization and root-finding with user-specified functions, providing an extensible class hierarchy that allows developers to readily implement new algorithms or interface to existing C and C++ routines.

S. Matsuoka et al. (Eds.): ISCOPE'99, LNCS 1732, pp. 154–163, 1999.
© Springer-Verlag Berlin Heidelberg 1999

2.1   The Interface

The primary method of the top level interface is solve(), which in an optimizer class will attempt to find the minimum of the specified merit function and, in a solver class, will attempt to find a multidimensional zero of the specified vector-valued function. The solve() method calls reset(), which returns the optimizer or solver object to its initial state, and complete(), which can be called in place of solve() if the user does not wish the object to be reinitialized:

```
void solve() throw(TxOptSlvExcept) {
  reset();      // put optimizer object into an initial state
  complete();   // attempt to find minimum of the merit function
}
```

In the event of a bad function evaluation, failure to converge, or other problems, this method will throw an exception of type TxOptSlvExcept.

The next code segment shows how complete() loops over three basic methods: isSolved() checks whether the algorithm has converged, step() carries out one optimization or solver step, and prepareStep() handles any tasks required before the next step is taken. The results are stored by setResult(), whether or not the algorithm converges. If the maximum number of iterations is exceeded, then an exception is thrown.

```
void complete() throw (TxOptSlvExcept) {
  for (numIter=0; numIter<maxNumIter; numIter++) {
    try { step(); }                 // try one step of algorithm
      // pass any exceptions on to the handler
    catch (TxOptSlvExcept& osx) { handleExceptions(osx); }
    if ( isSolved() ) {             // has algorithm converged?
      setResult(); return;          // store good results & return
    }
    try { prepareStep(); }          // prepare for the next step
      // pass any exceptions on to the handler
    catch (TxOptSlvExcept& osx) { handleExceptions(osx); }
  } // maximum # of iterations was exceeded
  setResult();                      // store unsuccessful results
  throw TxOptSlvExcept("Failure to converge");
}
```

This interface works well for a wide variety of 1-D and multidimensional optimization and root-finding algorithms. While the solve() method is fixed, the other methods are virtual, so new algorithms can be added by overloading the virtual building-block methods. Existing implementations of algorithms can be included within this interface by overloading the complete() method to call the desired implementation, then overloading the building-block methods, to either throw an exception or else call some appropriate method of the implemented algorithm.

The use of exception handling is required for the OptSolve++ class libraries to operate as robust software components that can be integrated into C++ applications. This results in a performance penalty; however, the additional run-time overhead only becomes important if the individual steps of the optimization algorithm are not computationally intensive.

2.2   The Class Hierarchy

Figure 1 shows a simplified class diagram for the TxOptSlv class library. The classes are templated over both the argument type (ArgType) and the return type (RetType) of the function to be optimized. This allows for situations where, for example, double precision arguments are used to avoid round-off error, but a single precision value is returned, or perhaps ArgType could be binary, integer or complex, while RetType could be of floating point type. When ArgType and RetType are both of floating point type, it is expected that ArgType will be of equal or higher precision than RetType.

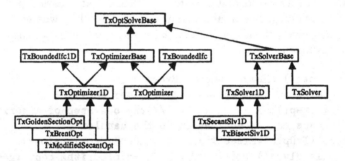

Fig. 1. Simple class diagram showing the abstract classes of the TxOptSlv class library, plus the concrete classes that implement 1-D algorithms.

From the TxOptSolveBase class, which defines the top-level interface described above, the hierarchy splits into optimizers and nonlinear solvers. From the TxOptimizerBase and TxSolverBase class, the two sub-hierarchies branch into 1-D and multidimensional classes. Figure 1 shows the 1-D algorithms that have been implemented at present. Figure 2 shows the multidimensional optimization algorithms that have been implemented. The multidimensional solvers are not discussed here, due to space limitations.

The TxOptSlv classes provide constructors that internally wrap the function to be optimized inside an appropriate TxFunc functor object. This approach allows users to instantiate an optimization object with a variety of function pointer types, while preserving the common interface provided by the functors. Users do not require any knowledge of the TxFunc library, if they can provide an appropriate C++ function pointer or pointers.

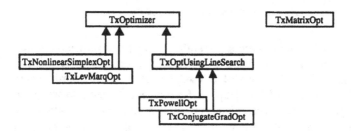

Fig. 2. Simple class diagram for the multidimensional nonlinear optimization classes of the TxOptSlv class library.

2.3 The Algorithms

A number of nonlinear optimization algorithms have already been implemented in TxOptSlv. The 1-D algorithms include golden section [1] and Brent [2], which do not need the function derivative, and a modified secant algorithm [3], which does require access to the derivative. The multidimensional algorithms include nonlinear simplex [4] and Powell [5], which do not need the gradient of the function, and the conjugate gradient method [1], [6], which does require the gradient. The TxOptimizer class provides an implementation of bound constraints (using a variable transformation approach), which is available to any unconstrained optimization class that inherits from it. General nonlinear constraints, using a penalty function approach [1], [6], [7], will be made available to all algorithms in the near future.

By default, calls for the derivative or gradient of a function will use finite difference methods. If the user supplies the constructor of an optimizer object with C++ function pointers that calculate derivatives as well as function values, then these supplied functions will be used in place of the finite differences.

We have also implemented algorithms for linear and nonlinear least squares. For the linear least squares problem,

$$min \, \|A * x - b\|_2 \, , \tag{1}$$

we have implemented methods using Cholesky, QR and singular value (SVD) decompositions [8], [9], which are encapsulated in the TxMatrixOpt class. The advantage of SVD is that is allows users to solve singular or close to singular systems of equations, as well as underdetermined problems. For nonlinear least squares,

$$min \, \|f(x)\|_2 \, , \; f(x) = 0.5 * \sum r_i^2(x) \tag{2}$$

we implemented the Levenberg-Marquardt algorithm, including quadratic constraints, using either Cholesky decomposition or SVD. [10], [11]

3   The TxFunc Library of Functors

A functor object represents a generalized function, returning a unique value associated with a set of input arguments. Functors can also contain auxiliary data or methods. This concept is used in the C++ standard library (see e.g. Sec.'s 11.9 and 18.4 of Ref. [12]). A functor can be used to wrap one or more function pointers, and it can be used to scale or to place constraints on a wrapped function. Most generally, a functor object can wrap an arbitrary sequence of calls to user-specified libraries, in order to calculate a value associated with the input arguments.

Figure 3 shows a simplified class diagram for the TxFunc class library. The abstract base class TxFunctorBase defines the fundamental interface for a functor, which consists primarily of the parentheses operator, (), declared pure virtual and templated over both the argument type (ArgType) and the return type (RetType). The hierarchy then branches into 1-D and N-D functors, very much like the optimizer and solver subhierarchies shown in Fig. 1. Some of the concrete classes, which wrap specific types of function or functor pointers, are also shown in Fig. 3.

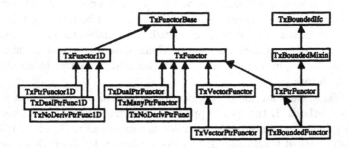

Fig. 3. Simple class diagram for the TxFunc class library, showing the abstract 1-D and multidimensional functor classes, as well as some of the concrete classes.

The N-D functors take a 1-D array of N values as the argument to the parentheses operator, with the array type specified by a template parameter. Making the parentheses operator a virtual function allows the functor classes and, hence, also the optimizer classes to be used polymorphically, which provides tremendous flexibility and extensibility. This approach does lead to a run-time performance penalty, because each function call now requires a number of table lookups; however, the resulting overhead is not important if the function is expensive to calculate. For example, in many real world applications, calculating the merit function involves running a design simulation.

For N-D functors with no specified derivatives, the TxNoDerivPtrFunc class defines an internal function pointer through a template parameter, FunPtrType, which must be a pointer to a C++ function, a TxFunctor or some other object with the parentheses operator () is appropriately defined. The TxDualPtrFunctor and TxManyPtrFunctor classes wrap different combinations of C++ function pointers in order to represent both the function value and its gradient.

The TxPtrFunctor class, also shown in Fig. 3, wraps a function or functor pointer that can simultaneously calculate the full gradient and return the function value. This class is also used as an intermediate base class for all functors that apply a variable transformation to the argument. For example, the TxBoundedFunctor uses a variable transformation to impose bound constraints on the argument of the interior functor. This functor implements the TxBoundedIfc interface class, which is also implemented by the TxOptimizer class, as shown above in Fig. 1.

Figure 3 also shows the vector functors: TxVectorFunctor encapsulates a function taking an N-D argument and returning a vector of M values. It implements a method for calculating the approximate Jacobian using the Broyden formula and has a pure virtual method for calculating the exact Jacobian. Since it derives from TxFunctor, it inherits an interface for a function taking an N-D argument and returning a single value. This parentheses operator () is implemented to calculate the sum of squares of the return values. The TxVectorPtrFunctor class holds an array of TxFunctor pointers, providing a concrete implementation of some TxVectorFunctor methods and allowing for direct access to each of the individual functions.

4 Template-Based Traits Mechanisms

The template-based traits mechanism [13], [14] allows C++ developers to write code for a templated class that refers to traits of the various numeric types in a very general way, making it possible to code numerical algorithms with one piece of source code that works equally well for types float, complex, double and so on. In OptSolve++ traits are implemented via templated structs. The upper portion of Fig. 4 shows how various unary traits have been defined for double and complex types. The lower portion of the figure shows how the binary trait PromoteType (result of combining two different numeric types in a binary arithmetic operation) has been specified for the combination of float and double types.

The traits idea and mechanism can be extended to create container-free C++ libraries [15], which allow the user to specify what sort of C++ container class the library must use. As for the traits of numeric types, container traits must be specified in a header file. The use of 1-D array classes in OptSolve++ is container-free in this sense, so the user is free to choose the type of the array argument taken by the multidimensional functor, optimizer and solver classes.

As shown in Fig. 5, container traits in OptSolve++ are implemented via templated structs. The upper box uses an std::vector as an example, showing that any container is expected to know the numeric type of the objects it contains and is also expected to be able to report its size and to resize itself. Also, containers are expected to allow clients to get and set individual elements through the

160   David L. Bruhwiler et al.

```
struct TxUnaryNumberTraits<double>
```
```
typedef double RealType;
typedef double AbsType;
typedef complex<double> ComplexType;
static double Norm(double x) {return x*x;}
static double Abs(double x) return abs(x);}
static double Real(double x){return x;}
static double Imag(double x){return 0;}
static inline AbsType Epsilon() {return AbsType(DBL_EPSILON);}
```

```
struct TxUnaryNumberTraits<complex<double> >
```
```
typedef double RealType;
typedef double AbsType;
typedef complex<double> ComplexType;
static double Norm(complex<double> x){ return norm(x);}
static double Abs(complex<double> x){ return sqrt(norm(x));}
static double Real(complex<double> x){ return real(x);}
static double Imag(complex<double> x){ return imag(x);}
static inline AbsType Epsilon() {return AbsType(DBL_EPSILON);}
```

```
struct TxBinaryNumberTraits<float, double>
```
```
typedef double PromoteType
```

Fig. 4. Examples of the OptSolve++ implementation of unary (above) and binary (below) traits, allowing for flexible templating over numeric types.

square brackets operator. The use of partial specialization allows these traits to be defined while leaving the ArgType template parameter completely general. The lower box in Fig. 5 shows how the binary numeric traits shown in Fig. 4 can be used to define binary container traits.

```
struct TxUnaryContainerTraits<std::vector<ArgType> >
```
```
typedef ArgType ValueType;
static void setSize(std::vector<ArgType>& vec, unsigned int dim)
    { vec.resize(dim);  }
static unsigned int getSize(std::vector<ArgType>& vec) const
    { return vec.size();  }
```

```
struct TxBinaryContainerTraits<std::vector<ArgType>,RetType>
```
```
typedef typename TxBinaryNumberTraits<ArgType, RetType> ::
                  PromoteType   PromoteType;
typedef std::vector<PromoteType>  PromoteVectorType;
```

Fig. 5. Examples of OptSolve++ implementation of unary (above) and binary (below) container traits, making the TxOptSlv and TxFunc components container free.

5   The TxLin Linear Algebra Library

Linear algebra is a cornerstone for mathematical libraries. OptSolve++ comes with a linear algebra component, TxLin, which is required by some of the optimization and solver algorithms. TxLin uses reference counting [16], which allows for efficient handling of large matrices by allowing multiple instances of matrices with the same data to share the same region of memory for as long as possible. The public interface is presented by the envelope classes, named with a Tx prefix, while the implementation is hidden in the parallel hierarchy of letter classes, named with a Pr (private) prefix.

Figure 6 shows the parallel letter and envelope hierarchies for TxLin. The underlying reference counting mechanism is set up in the Pr/TxRefCount classes, while the Pr/TxTensor classes define a general tensor class that is templated over both the dimension and the numeric type of the elements. The Pr/TxVector classes specialize to the case of a 1-D tensor, and the Pr/TxMatrix classes specialize to the case of a 2-D tensor – in both cases adding relevant linear algebra capabilities, while the Pr/TxSquareMatrix classes further specialize to the case of square matrices and include the appropriately specialized algorithms. The templated TxLin vector and matrix classes overload the usual mathematical operators for matrix and vector arithmetic, and they implement many of the standard linear algebra operations, with the same source code being used for float, double and long double precision.

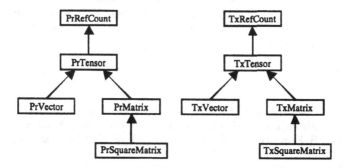

Fig. 6. Simple class hierarchy for the reference-counted TxLin linear algebra library.

At present, TxOptSlv explicitly uses TxLin. Future versions of OptSolve++ will abstract this connection, so that users can replace TxLin with other C++ linear algebra class libraries. This abstraction can be accomplished through the container-free techniques described above, although this becomes syntactically very awkward and is also dangerous in that one must then assume a great deal regarding the capabilities of the other libraries that might be used. Thus, we see a strong need for some degree of standardization in both the interface and the minimum capability of C++ linear algebra components. There has been some informal discussion of possible standards [17], but no consensus has yet emerged.

6   Run-Time and Compilation Performance Issues

OptSolve++ has been used to solve standard test problems of low to moderate dimensionality from the literature [18] and also in particle accelerator design applications. The implemented algorithms work well. Techniques such as template meta-programming, which can yield improved performance, were avoided so that more C++ compilers could be supported. OptSolve++ supports a wide array of compilers [19], making it truly cross-platform software. Heavy use of templates can also make compile times so slow that development is very difficult. In OptSolve++, the bulk of the implementation is kept out of header files, so that compile times are fast enough to do rapid development work. This approach requires various template instances to be explicitly instantiated for all classes, but the reduction in flexibility yields a dramatically improved development environment.

Acknowledgements

The authors wish to thank D. Alexander and K. Luetkemeyer of Tech-X Corporation for discussions regarding design and interface, for contributions to early versions of the library, for assistance in software testing and in porting to new platforms, and for comments on drafts of this paper. This work is supported by the U.S. Department of Energy, under grant number DE-FG03-96ER82292, and by Tech-X Corporation.

References

1. E. Polak, Computational Methods in Optimization, (Academic Press, 1971).
2. R. Brent, Algorithms for Minimization without Derivatives, (Prentice-Hall, 1973).
3. W. Press, B. Flannery, S. Teukolsky and W. Vetterling, Numerical Recipes, (Cambridge University Press, 1989).
4. J. Nelder and R. Mead, Computer Journal 7 (1965), p. 308.
5. D. Himmelblau, Applied Nonlinear Programming, (McGraw-Hill, 1972), p. 167.
6. D. Himmelblau, Applied Nonlinear Programming, (McGraw-Hill, 1972), p. 167. R. Fletcher, Practical Methods of Optimization, 2nd edition (John Wiley and Sons, 1987).
7. J. Moré and S. Wright, Optimization Software Guide, (SIAM, 1993).
8. L. Trefethen and D. Bau, III, Numerical Linear Algebra, (SIAM 1997).
9. J. Stoer and R. Bulirsch, Introduction to Numerical Analysis, (Springer-Verlag, 1980).
10. A. Bjorck, Numerical Methods for Least Square Problems, (SIAM 1996).
11. J. Moré, The Levenberg-Marquardt Algorithm: Implementation and Theory, in Numerical Analysis, Lecture Notes in Mathematics 630 (Springer-Verlag, 1978), p. 105.

12. B. Stroustrup, The C++ Programming Language, 3rd edition (Addison-Wesley, 1997).
13. N. C. Meyers, Traits: a New and Useful Template Technique, C++ Report 7 (June, 1995).
14. T. Veldhuizen, Using C++ Trait Classes for Scientific Computing, (March, 1996), available on-line at URL http://oonumerics.org/blitz/traits.html.
15. G. Furnish, Container-Free Numerical Algorithms in C++, Computers in Physics 12 (May, 1998).
16. S. Meyers, More Effective C++, 2nd edition (Addison-Wesley, 1998).
17. The html archive for the OONSTD (Object-Oriented Numerics Standards) mailing list, on-line at URL http://oonumerics.org/oon/oonstd/archive.
18. J. Moré et al., ACM Trans. Math. Soft., Vol. 7 (1981), p. 17.
19. A complete list of supported operating systems and compilers can be found on the web at URL http://www.techxhome.com/products/optsolve.

Are Generic Parallel Algorithms Feasible for Quantum Lattice Models?

Matthias Troyer

ETH Zürich, Zürich, Switzerland

Abstract. The investigation of strong quantum effects is one of the major research areas in physics with important application aspects, like the high temperature superconductors. The only reliable approaches to these systems are large scale numerical simulations, made difficult by the exponential scaling of most algorithms. We present efforts in developing a generic and parallel program package for the simulation of such systems using generic programming techniques of the C++ language, including expression template techniques for quantum operators.

1 Introduction

Electron correlation effects in solids, caused by strong quantum effects are the origin for a number of interesting phenomena – like high temperature superconductivity – with potentially very important technical consequences. Analytic as well as quantum chemistry approaches fail for these materials since they rely on quantum effects being small. In order to calculate the properties of such materials the full quantum effects must be treated! This makes the simulations exponentially hard, typically scaling like 4^N.

A variety of algorithms have been developed, each of which is useful for the calculation of some properties of certain models. The programs are usually written and optimized in Fortran specifically for one model, which contrasts sharply with the generality of the algorithms. In this paper we report on the PALM++ project (Parallel Algorithms for Lattice Models), an effort to implement a generic and parallel toolkit for these algorithms using generic and object oriented programming features of C++, including expression template techniques for quantum operators.

2 Quantum Lattice Models and Bit String Representations

2.1 Bit String Representation of Quantum States

As the full Schrödinger equation for a material cannot be solved, simplifications have to be made. In contrast to quantum chemistry, which simplifies the interaction we treat the *full* quantum problem but reduce the atoms to one or a few relevant orbitals. Different effective models are thus derived for various

S. Matsuoka et al. (Eds.): ISCOPE'99, LNCS 1732, pp. 164–169, 1999.

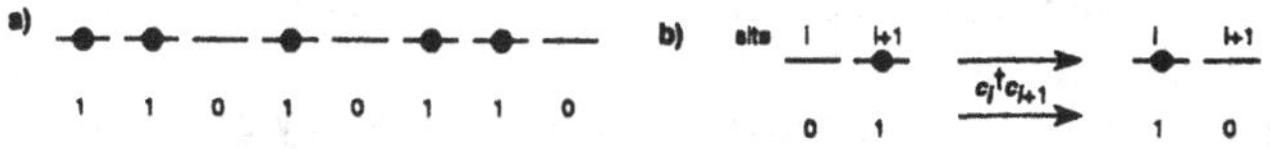

Fig. 1. a) A quantum mechanical state for spinless particles and its representation as a bit string; b) The operation of the quantum operator $c_i^\dagger c_{i+1}$ on a quantum state and on the bit string representation.

materials, but they all share common properties which can be used for generic algorithms.

The Hilbert space of the model is the vector space spanned by all possible occupation configurations of these orbitals. These occupations are most naturally represented as bit-strings, where in the simplest case a "1"-bit represents the occupation of an orbital, as is shown in Fig. 1a for a simple model with at most one particle per orbital. For more complex models two or more bits have to be used to represent the state of an orbital.

2.2   Quantum Operators

The Schrödinger equation is a linear eigenvalue equation $H|\psi\rangle = E|\psi\rangle$. The Hamilton (energy) operator H is a sparse Hermitian matrix, describing the physics of the quantum model. Its eigenvalues E are the energy levels, and the eigenvectors the wave functions $|\psi\rangle$, from which physical properties can be calculated. This Hamilton operator is usually written in the second quantized form using operator notation. For a chain of our simple non-interacting spinless particles – spinless fermions or hardcore bosons for the experts – the Hamilton operator is

$$H = -t \sum_{i=0}^{L-1} c_i^\dagger c_{i+1} + c_{i+1}^\dagger c_i \tag{1}$$

The symbol $c_i^\dagger$ is an operator which adds a particle on site i. The adjoint operator c_i removes a particle from site i. The product $c_i^\dagger c_{i+1}$ thus attempts to hop a particle from site $i + 1$ to site i, as is illustrated in Fig. 1b. If this is possible there is a matrix element $-t$ connecting the two states.

$c_i^\dagger c_{i+1}$ is thus a very sparse matrix with at most one nonzero matrix element per row! The total Hamilton operator, a sum of $O(N)$ such terms and thus also extremely sparse, with dimension scaling like $O(\exp(N))$ but only $O(N)$ nonzero matrix elements per row. The matrix elements can be calculated using bit operations, as will be discussed below.

2.3   Calculating Matrix Elements

H is blockdiagonal, consisting of blocks for each particle number N. The basis for the N particle subspace of our simple model consists of all bit strings of length L, with N bits set to one. These bit strings are usually stored in an array, which we will call **state**.

To obtain the matrix elements of H we need to apply all the operator terms in H to each of the states. For a term like $-t c_i^\dagger c_{i+1}$ (see Fig. 1b) this is done as follows:

> loop over all indices $n = 1 \ldots$ dimension of basis
> > apply the quantum operator c_{i+1} by clearing bit $i + 1$
> > > $s' \leftarrow state(n) \& \tilde{\ }(1 << (i+1))$
> > apply the quantum operator $c_i^\dagger$ by setting bit i
> > > $s'' \leftarrow s' | (1 << i)$
> > if s'' is one of the basis elements
> > > determine the index m such that $s'' = \textbf{state}(m)$
> > > set the matrix element $H_{m,n} \leftarrow -t$

The bit operations can be optimized by implementing the sum $c_i^\dagger c_{i+1} + c_{i+1}^\dagger c_i$ by an **exclusive or** with a bit mask that has bits i and $i + 1$ set.

The time consuming part is the determination of the index m of a state s for which perfect hashing schemes are preferred. For memory reasons multi-step hashing are used for most models. Translational symmetry can be used to further reduce the number of basis states. Translations are again performed by bit operations (in this one-dimensional example using circular bit shifts).

3   Algorithms for Quantum Lattice Models and their Parallelization

Several types of algorithms, which complement each other, have been developed for the simulation of quantum lattice models.

1. Exact diagonalization. Although the size of the Hilbert space grows exponentially like $O(\exp(L))$ the most interesting extreme eigenvalues of such a matrix can be calculated for systems with up to $L \approx 36$ orbitals using iterative methods, like the Lanczos algorithm [1]. Parallelization can be done by parallelizing the sparse matrix-vector multiplications. As the matrices are often too large to be stored the matrix elements have to be calculated efficiently "on-the-fly" at each iteration.

2. Series expansion techniques express the properties of *infinitely large* systems as a power series in a small parameter, such as the inverse temperature $1/T$: $A = \sum_n a_n T^{-n}$ The calculation of the coefficient a_n requires calculating $\mathrm{Tr} H^n$. H^n is calculated as in exact diagonalization. The trace $\mathrm{Tr} H^n$ is calculated by looping over all basis vector and can be trivially parallelized.

3. Quantum Monte Carlo methods map the quantum system to a classical system in a higher dimensional space, which is then simulated using standard Monte Carlo methods. Monte Carlo algorithms usually scale polynomially with the number of lattice sites N. For most models of electrons however, there is a severe cancellation problem, the so-called "negative sign problem" , which again causes exponential scaling of the algorithm. Parallelization of Monte Carlo simulations is straight-forward and was discussed last year at this conference [2].

4. The Density Matrix Renormalization Group method [3] is a recent development, which reduces the exponentially large number of basis states by using just a small number $M = m^2$ of important basis states, which are determined by iteratively growing the lattice. Numerically, the Hamiltonian matrix H in this truncated basis is no longer sparse, but still block-sparse and a matrix-vector multiplication can be parallelized by distributing the blocks over the nodes of a parallel computer. We need to investigate whether the POOMA-II array library [4] can be used for these purposes. The DMRG method scales only polynomially in the number of orbitals N and basis states m and is thus ideal in one dimension. In higher dimensions the number of states m needed for accurate results grows exponentially with N.

4   Are Generic Algorithms Feasible?

Due to the exponential scaling of most algorithms parallel implementations are essential to obtain good accuracy. The aim of the PALM++ project is to develop parallel and generic programs for the four algorithms discussed above. Here we wish to discuss whether a generic implementation is feasible. Of special concern are three issues:

- Are bit operations in C++ as efficient as their FORTRAN counterparts?
- Can these algorithms be implemented in a generic and still efficient way?
- Can the implementation of a model be simplified using expression templates for quantum operators, similar to the expression templates technique for array operations developed by Todd Veldhuizen [5]?

The first question has been positively answered by our benchmarks. However, independent of the language, compiler and hardware support for bit counts can be essential for optimal codes for fermionic models. Bit counts are used extensively in cryptographic decoding sieves, hence e.g. Cray vector supercomputers have good hardware support. On workstations without such hardware support bit counts need to be done using about a dozen bit operations. Then the CPU performance and not memory bandwidth limit the performance.

The remaining two questions will be answered in the next sections.

4.1   Generic Algorithms for Classical Simulations

Generic algorithms, as used in the standard template library, are well known to be similarly efficient as specialized algorithms. For array operations the Blitz++ library[6] is also on average about as efficient as optimized Fortran programs. Many physics simulations however are not based on simple array operations, but use more complex data structures like stacks and trees.

As a benchmark case for such a simulation we have chosen the Wolff cluster algorithm for classical Monte carlo simulations. This algorithms was originally derived in a generic form [7], but until now only implemented for specific models. We have implemented this algorithm generically [8], partially using the standard template library and compared the performance with an optimized Fortran implementation for a classical Heisenberg spin model. For this model the classical spins are represented by N-vectors, which we implement using the TinyVector<double,N> class of the Blitz++ library [6].

We find that the codes perform comparably. Which version is faster depends on the platforms and compilers used. On a 20^3 lattice at the critical temperature, the generic C++ code was 23% slower than the specialized Fortran code on a 333 MHz UltraSparc 5, but 33% faster on an IBM SP-2. As C++ compiler the KAI C++ compiler version 3.3e with optimization options +K3 -O3 was used. As fortran compiler the vendor supplied native Fortran 77 compiler was used, with highest optimizations. More details, including source codes, will be published in the near future and can be found on the web page http://www.itp.phys.ethz.ch/compphys/.

4.2 Expression Templates for Quantum Operators

The bit string representation of quantum states and operators is very efficient but it is a big task to implement a new operator or model. Ideally we want to be able to code an operation such as $|\phi\rangle = (-tc_i^\dagger c_j + \mu(n_i + n_j) + V n_i n_j)|\psi\rangle$ as phi = (-t*cdag(i)*c(j) + mu*(n(i)+n(j)) + V*n(i)*n(j))*psi.

We have developed an expression template (ET) implemntation for the hardcore boson models and plan to extend it to more complex models in the near future. This implementation has identified the following issues that an ET implementation for operators has to address in addition to standard ET techniques:

- In C++ operators bind from left to right, while in the standard quantum mechanical notations operators act to the wave function on the right. cdag(i)*c(j)*psi is parsed as (cdag(i)*c(j))*psi, but this has to be rearranged to cdag(i)*(c(j)*psi) before evaluating it, as we want to be able to use the standard notation and not force the user to set extra parentheses. The implementation can only be outlined here for space reasons. A more detailed discussion and the source code is available at http://itp.phys.ethz.ch/compphys/qet/

  1. An expression like phi=A*B*psi is parsed by the compiler and an expression object for (A*B)*psi is constructed using standard ET techniques.
  2. In the assignment the expression is evaluated for all basis states s. For this a member function of the expression object evaluate(s) is called.
  3. This function then calls a member function evaluate(psi,s) of the expression object for A*B, which in turn evaluates the rearranged expression A*(B*psi).

- In contrast to array operations, where functions are performed elementwise on arrays, the quantum operators are actually sparse matrices, and the quantum ET implementation performs a sparse matrix-vector multiplication. This can be done easily in our implementation, as e.g. when the expression A*(B*psi) is evaluated, the subexpression (B*psi) can easily be evaluated for more than one state.

Note that similar problems have to be dealt with if one wishes to implement stencil operators or matrix multiplications using ETs. An example are array stencils that allow $(v \cdot \nabla)u$ to be coded as (v*grad)*u.

- There are many tricks that can speed up these operator expression. One that was mentioned in Sect. 2.3 is the optimized evaluation of $c_i^\dagger c_{i+1} + c_{i+1}^\dagger c_i$ by one "exclusive or" operation. These optimizations are implemented using partial template specialization.
- The vectors phi and psi can have vastly different dimensions. In such cases it is better to apply the adjoints of the operators to phi. Code can be generated for both versions and the decision on how to evaluate the expression taken at runtime.

5   Conclusions and Outlook

A combination of object oriented and generic programming techniques, including extension of the expression template ideas to quantum operators allows the implementation of generic programs for quantum lattice models. The PALM++ project (Parallel Algorithms for Lattice models) is an effort lead by the author, which aims at developing a generic parallel toolkit for such simulations. These parallel programs will allow new models to be investigated with minimal effort, without requiring several weeks to months of coding .

C++ is at the moment the only language which permits generic but still optimal implementations. The effort involved in programming expression templates makes it clear that C++ is by no means the ideal language for these efforts. We have identified some of the major problems encountered and hope that computer scientists will develop languages that are better suited for generic scientific computing. Bit operations in C++ were found to be efficient, with the exception of bit counts, which need to be supported by hardware and compilers to be efficient.

References

1. C. Lanczos, J. Res. Natl. Bur. Stand. **45**, 225 (1950).
2. M. Troyer, B. Ammon and E. Heeb, in the Proceedings of ISCOPE '98, Lecture Notes in Computer Science **1505**, (Springer Verlag, 1998).
3. S. R. White, Phys. Rev. Lett. 69, 2863 (1992).
4. S. Karmesin et al. in the Proceedings of ISCOPE '98, Lecture Notes in Computer Science **1505**, (Springer Verlag, 1998).
5. T. Veldhuizen, C++ Report **7**, 26(1995); ibid. **7**, 36(1995).
6. T. Veldhuizen in the Proceedings of ISCOPE '98, Lecture Notes in Computer Science **1505**, (Springer Verlag, 1998).
7. U. Wolff, Phys. Rev. Lett **62**, 361 (1989).
8. E. Heeb and M. Troyer, Bull. Am. Phys. Soc. **44**, 1, YB13.07.

A Runtime Monitoring Framework for the TAU Profiling System

Timothy J. Sheehan, Allen D. Malony, and Sameer S. Shende

University of Oregon, Eugene OR, USA

Abstract. Applications executing on complex computational systems provide a challenge for the development of runtime performance monitoring software. We discuss a computational model, application monitoring, data access models, and profiler functionality. We define data consistency within and across threads as well as across contexts and nodes. We describe the TAU runtime monitoring framework which enables on-demand, low-interference data access to TAU profile data and provides the flexibility to enforce data consistency at the thread, context or node level. We present an example of a Java-based runtime performance monitor utilizing the framework.

1 Introduction

The building of complex software systems is increasingly targeted to high- performance computing architectures that support thread-based parallel execution within a shared memory context and process-based (component-based) distributed execution across multiple physical nodes [1,2,3]. Object-oriented computing is particularly suited to such architectures because it can naturally capture thread abstractions and extend the object interaction paradigm to distributed environments [1,4,5]. While rich programming frameworks and layered middleware systems help tame the complexity of software built for next-generation high-performance environments [6,7], the observation of program operation and performance is critical to understanding and improving code efficiency.

Most monitoring systems built for parallel system environments are created for a specific purpose (e.g., performance measurement [8], debugging [9], or computational steering [10]), and with a specific computational model in mind. As a result, the monitoring system design and implementation tends to reflect decisions concerning the requirements for system observation, the type and level of instrumentation, the online accessibility to monitored information, the degree of consistency with respect to system state, and the acceptability of monitoring costs. It is difficult to construct a *monitoring framework* that is flexible enough to apply in different computational contexts and in which monitoring parameters can be selectively controlled. It is not simply an issue of monitor portability. The framework should support the construction of monitor software that operates in a manner consistent with the parallel software paradigm and its execution without excessively imposing it own constraints.

S. Matsuoka et al. (Eds.): ISCOPE'99, LNCS 1732, pp. 170–181, 1999.

In this paper, we describe the TAU monitoring framework and its use in observing the performance of parallel applications built for complex, high- performance computing architectures. The framework instantiates a monitoring system that augments the application and system software with monitor interaction components and distributed monitor communication and control mechanisms. This monitor infrastructure is intended to interface with different online instrumentation facilities, extending the range of monitor applications. Integrating monitoring capabilities with complex software requires knowledge of the monitor's influence on program execution, and the monitoring framework should support a variety of alternative methods of data access. We discuss monitor interaction models and certain problems that can arise in the context of online access to performance data provided by the TAU profile library.

2   Model of Computation

A monitoring model must be based on an underlying computational model which must, in turn, reflect the real computing environment. The computational model we are targeting is illustrated in figure 1. A *node* is a physically distinct entity with one or more processors. A node may link to other nodes via a protocol-based interconnect, ranging from proprietary networks, as found in traditional MPPs, to local- or global-area networks (e.g. the Internet). A *context* is a distinct virtual address space residing within a node. Multiple contexts may exist on a single node. Multiple *threads* of execution, both user and system level, may exist within a single context. Threads within a context share the same address space.

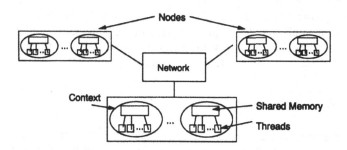

Fig. 1. The HPC++ computational model.

This computational model is general enough to apply to many high- performance architectures as well as to different parallel programming paradigms. This enables us to consider a monitoring framework with respect to different computational model views. In addition, the model is the basis for HPC++ [1], a high-performance parallel and distributed object-oriented system that supports both task- and data-parallel programming paradigms.

3   Monitoring Systems

Monitors can be used for a wide variety of functions including data visualization, performance analysis, computational steering, and debugging. All of these activities utilize at least one of two basic operations on a executing application: the observation of the program's data and interaction with the executing code. Implicit in this description is a computational entity that provides access to the program's context(s) and mechanisms to involve itself with the executing code. We term such an entity a *monitor agent*. Also implicit is a *monitor client* which directs the observation and interaction.

3.1   Monitor Agent

In the HPC++ model of computation, the monitor agent must reside within the executing application's context(s) in order to access its memory. This agent may be a *direct (active) agent* or an *indirect (passive) agent*. A direct agent is invoked from code instrumentation to provide data access to the monitor client. An indirect agent, on the other hand, runs in a separate thread of execution, with full access to the context memory. Additional software resources are not needed to use a direct agent, but execution of the computation is directly impacted. An indirect agent can observe passively with less interference on the application's thread(s) of execution.

3.2   Monitor Client

The monitor client communicates with the monitor agent, directing the agent's actions and interpreting the monitored information. A client may reside within the same context as the agent or in a separate context. Normally, to avoid competing for the application's computational resources, the client resides outside of the context of the monitor agent and communicates with it in the same manner that other threads communicate across contexts.

4   Profiler Structure

A *profiler* keeps summary statistics of application execution based on the occurence of events[11]. Profiling systems generally track two different types of entities: user defined events and performance of blocks of code such as a routine or a group of related statements. For purposes of this paper we term such code blocks *functions*. The profiler maintains a database of user defined event counters. Whenever a user defined event is triggered, its counter is incremented. Count and execution time values are profiled for functions. Time is measured by elapsed wallclock time, or can be substituted by hardware performance counters to measure low-level CPU activity such as secondary data cache misses, or instructions issued. The profiler maintains a database of function information and an image of the callstack. The database includes the time consumed by each profiled function. The callstack image includes the starting time of each profiled function called. When a function completes and its entry is popped from the top of the callstack image, its starting time is subtracted from the current system time and this is added to the cumulative time for that function's database entry. At the end of a run, when the callstack image is emptied and the database has been updated, the data is written to a file where it can be accessed for post mortem processing and visualization. The profiler data structures are maintained on a per thread basis, and executing threads update their profiling data completely independent of one another.

5   System Snapshot

Post mortem profile analyses are derived from the data monitored during execution. A runtime monitor should provide similar views of the system while the application is executing. Such a view can be derived from the callstack image and count and time databases maintained by the profiler. We term this a *snapshot*, meaning a view of the state of the system at a given point in time.

5.1   A Consistent Snapshot

The concept of consistency is important in the representation of the state of the system. In post mortem analysis, the state of the system is simply the static state at program termination. At runtime, however, the system state is constantly evolving and its consistency must be taken into consideration when obtaining profile data.

If the monitor attempts to read the profile database or callstack while it is being updated by a thread, the information may not be consistent and the snapshot of the system may be erroneous. Thus, for a thread, a consistent snapshot requires that the acquisition of all the profile data be an atomic operation. This is accomplished by locking all of a thread's profile data before reading it.

The definition of a consistent snapshot can easily be extended to a context. Within a context a consistent snapshot is a set of images, one per thread, which represent the state of all the threads in the context at the same point in time. Acquiring a consistent snapshot within a context requires that the profile data for each thread be simultaneously locked and that a lock only be released once its thread's data is read. Likewise, a consistent snapshot of a multi-context system requires the simultaneous locking of data for all threads in all contexts, and the releasing of any lock only after its data is read.

5.2   Practical Considerations

As mentioned above, a truly consistent snapshot of multiple threads requires simultaneously locking the access of all threads. In most systems, this can only be done to the resolution of time it takes the lock command to reach all threads. Simultaneous locking could have a profound effect on application performance. While this doesn't stop the execution of the application, subsequent function entry and exit operation and event counter operation is blocked until the monitor agent has obtained the performance data. The possible performance impact imposed by absolute consistency may be justified when debugging a program for problems such as deadlock or device access conflicts.

A less costly alternative to an absolutely consistent snapshot is an approximately consistent snapshot. This is often termed *loose consistency*. One implementation of this is effectively looping over each thread, locking its profile data, reading its profile data, then releasing the lock. What is produced is analogous to a radial sweep radar image. The only portion of the display that is absolutely current is where the sweep is right now. The whole image, however, is approximately current. The advantage of this method is that it only has the potential to affect the execution of one process at a given time, and the maximum effect on a process is the time it takes to read that process's callstack and database images. Loose or approximate consistency is desirable when the one of the goals of the monitoring system is to minimize the extent of intrusion, for example with real-time performance observation.

6   Runtime Data Access

Accessing data at runtime involves trade-offs between how current the data is, the timing of access, and the impact of the access on the application. We consider three passive agent models for runtime monitoring of application profile state: *push*, *push-pull*, and *pull*.

6.1   Push Model

Figure 2(A) shows the push model of data access to profile data [12,13]. During execution, the callstack and profile database are updated. Periodically, the application pushes current profile data into a block of memory accessible to the monitor agent, a separately executing thread. The application signals the monitor agent that the data is available. *Synchronous* access by the monitor requires the application thread signalling the monitor to block until the monitor agent signals that data access is complete. If access to the data is *asynchronous*, the application thread continues while the monitor operates, later blocking until data access is finished. The application controls where and when state can be observed.

The advantage of the push model is that instrumentation in the profiler controls where and how profile data can be accessed, and when it is made available to the monitor agent. It is possible to maintain better consistency in this case, but can be intrusive on execution.

6.2   Push-Pull Model

The push-pull model is shown in figure 2(B) [14]. In this model, profile data is periodically pushed into a profile data buffer. The monitor agent has constant

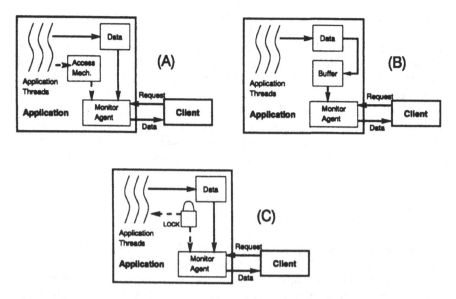

Fig. 2. Models of data access: (A) push model; (B) push-pull model; (C) pull model.

access to the the buffer and can pull data from it any time the application is not writing to the buffer. This solves two of the problems with the simple push model. First, the application thread does not have to block while the monitor agent accesses data. Second, the monitor agent can access the data in the buffer any time the application is not writing to the buffer. However, the data in the buffer does not reflect changes in the data as the program runs.

6.3   Pull Model

The final model is the pull model as shown in figure 2(C). In this model both the application thread and the monitor agent have access to the profile data at all times. When the application thread writes, or the monitor agent reads, a lock is used, insuring that the data does not change during the read. The disadvantage of this method is that while the monitor agent has the lock during a read, the application may block on a write to the database, impacting program performance.

The pull model does not provide for notification of the monitor agent when the application thread completes. In order for the agent to gracefully handle application shutdown, the application must set a variable indicating it is ready to shut down, then block waiting for a signal from the monitor. When the monitor detects that the application has finished, it can react appropriately then signal the application to terminate.

An application thread's operation on profile data (lock, access, release) is atomic and involves no communication with other application threads. A thread cannot be blocked waiting for a message while it holds the lock on the profile data. Thus a monitor agent that locks profile data - even for all threads - cannot introduce deadlock within an application unless it fails to release one or more of the threads. Furthermore, since the monitor agent runs as a separate thread from the application, it can access performance data even if the application itself is deadlocked.

This framework allows multiple monitor agents to access performance data from the same application. The model does not preclude deadlock caused by multiple agents simultaneously attempting to obtain locks from multiple threads. It is the responsibility of the framework user to insure that deadlock is not introduced in this manner.

7   TAU Runtime Monitoring Framework

We have have designed and implemented the *TAU runtime monitoring framework* that interfaces with the TAU profiling package [15,16]. The goals of this framework are threefold: provide a consistent snapshot of performance data within an executing application; allow access to performance data at any point during execution; and to impose the smallest possible penalty on application performance.

7.1   Architecture

This system is an implementation of that depicted in figure 2(C). Each monitor agent in an application context provides access to a monitor client running in another context. For applications running across multiple contexts, an agent is present in each context. Hence, a monitoring program can create and manage multiple monitor clients as a means to interact with agents in multiple contexts. Each client object can access its agent's data on demand.

HPC++ [1] is used for both the monitor agent and client; monitoring clients are implemented as objects. It provides remote function invocation (RMFI) as well as data transport through global pointers across contexts. RMFI allows the client to remotely lock and unlock the application's data and direct the monitor agent to gather data while global pointers provide the client with remote data access.

7.2   Functionality

The HPC++ monitor agent is not limited to communication with only one client object. This allows for multiple client applications in multiple contexts to access the data from a single application. This allows for *collaborative monitoring* in which users on several different systems can use the framework to simultaneously observe the performance behavior of an application executing on a separate system.

Also, because the client is an object, any number of them can be created by a single application, so a single monitoring application can spawn a client object for each context in an executing application and simultaneously display data from all contexts as illustrated in figure 3.

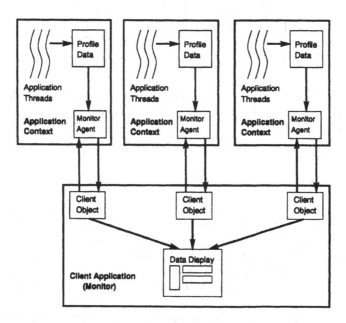

Fig. 3. Multiple clients accessing data from multiple application contexts.

7.3 Consistent Snapshots

This framework has the ability to lock the databases and callstack images in a context simultaneously and independently of when it reads the data. This provides the user with the ability to enforce consistency within a thread or among threads. In addition, all threads in all contexts can be locked at the same time within the limit of how long the system takes to issue one lock command for each context. Data can be read from each context before its thread is released, or from all contexts before any thread is released. The user can choose the tradeoff between the level of consistency of performance data and the potential performance penalty imposed on the executing application.

7.4 Error Detection and Recovery

The monitor agent can detect and recover from any application error that the application can trap. This is handled in the same manner as normal program

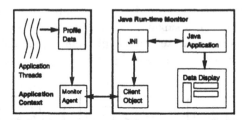

Fig. 4. Architecture of Java-based runtime monitor

termination. Exceptions originating in software layers upon which the framework is built, for example HPC++ and Nexus, can also be handled as long as they propagate up to the higher level function calls the framework uses. Development of robust error detection and reaction is ongoing.

8   Example Monitor Application

We have implemented a runtime TAU profile monitor based on the TAU monitoring framework. Shown in figure 4, the monitor uses a Java-based front end that interfaces with C++ through the Java Native Interface (JNI). This allows it to create a client object and attach to an executing application instrumented with TAU. At user-defined time intervals, the runtime monitor accesses application profile data through the monitor agent and displays profile data to the user.

Three different data displays are available. The *Exclusive Time* display (figure 5) provides a per thread overview of the exclusive time spent in each routine. A bar with contrasting colors graphically depicts the data, while a scrollable color key shows routines, exclusive times, and percentages. In the *Function Detail* display (figure 5), the functions from the database are shown for the selected thread. The function data is shown for the selected function. Also available is a callstack image.

9   Releated Work

The TAU runtime monitoring framework supports on-line access to distributed performance data through a high-level object-oriented interface that can be used to control the frequency, amount, and consistency of data monitored, without requiring clients to know the location of application components. These aspects of parallel and distributed monitoring have been the focus of recent attention in other research projects in debugging, performance analysis, adaptive control, and computational steering. We consider the OMIS/OCM, Autopilot, and MOSS projects.

The On-line Monitoring Interface Specification (OMIS) [17] and the OMIS compliant monitoring system (OCM) [18] target the problem of providing a universal interface between on-line tools and a monitoring system, which need not be

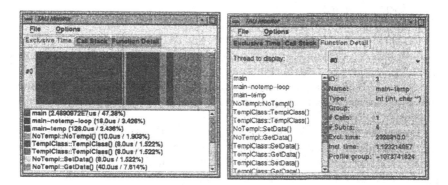

Fig. 5. TAU runtime monitor exclusive time and function detail displays.

targeted to a particular class of tools. OMIS supports an *event-action* paradigm for monitoring where events and their response actions are determined by tool requests to the monitoring system which provides event mapping and request distribution to different "objects" of a computation. OCM implements a distributed client-server system for these services, interconnecting local monitors with a centralized request and reply distributor process. Although sharing the same architectural objective and concern for asynchronous monitor operation, the TAU monitor differs in its use of a distributed object system for tool interaction with monitored components. The TAU monitor promotes a request-style of interaction (i.e., a pull model), but through more direct, high-level mechanisms.

Autopilot [19] extends the event-action paradigm to adaptive control of distributed application performance using *sensors* to extract performance data, *clients* for decision making, and *actuators* for application control. Autopilot's relation to our work is in its performance evaluation focus and its support for interconnecting sensors, clients, and actuators via global naming and remote services. However, the TAU monitor can access performance data directly via an object-oriented method interface, allowing client tools to decide dynamically the type of access to perform.

In contrast to event-stream-based monitoring, the Mirror Object Steering System (MOSS) [20] implements a model of monitoring and steering that treats application entities as objects with interactivity state and methods added to support distributed interactive data access and computation. The MOSS model and implementation approach is consistent with ours except the TAU monitor uses HPC++ as its distributed object system. Whereas MOSS is more extensive in its application scenarios, the TAU monitor concentrates on profile performance data. However, both employ object principles that leverage the natural use of existing synchronization code in a parallel and distributed application, as in the TAU profiling library.

10   Conclusions

Of the three monitoring models discussed, the TAU runtime monitoring framework is designed for the less-constraining, but more difficult to implement, pull model. Based on the general computation model in HPC++ and the portability of the TAU profiling system, a runtime monitor built from the framework can target complex monitoring needs for a diversity of parallel and distributed computing platforms.

Important features of the monitoring framework are its ability to support multiple levels of monitoring to capture consistent snapshots (single thread, multi-thread, or multi-context) and multiple monitor client interactions. This gives monitor developers the flexibility to build runtime monitor solutions specific to the observational and platform constraints. Additionally, the framework allows the developer to augment the monitor's functionality, such as to choose where monitored data analysis takes place, in the monitor agent or monitor client.

Our implementation of a Java-based runtime monitor based on this framework demonstrates its use for online access to TAU profiling data. We are presently pursuing the integration of this monitor in large-scale ASCI applications.

Acknowledgments

This work was supported by the Department of Energy DOE 2000 program (#DEFC0398ER259986) . We would like to thank the Los Alamos National Laboratory for their support. Matthew Sottile, and Chad Busche at the Univ. of Oregon, contributed to the implementation of the system.

References

1. Gannon, D., Beckman, P., Johnson, E., Green, T., Levine, M.: HPC++ and the HPC++LIB Toolkit, Technical Report Department of Computer Science, Indiana University (1998).
2. Laure, E., Mehrotra, P., Zima, H.: Opus: Heterogeneous Computing With Data Parallel Tasks, Technical Report TR 99-04, Institute for Software Technology and Parallel Systems, University of Vienna URL:http://www.par.univie.ac.at (1999).
3. OpenMP: OpenMP   Fortran   Interpretations   Versions   1.0, URL:http://www.openmp.org (1999).
4. Chandy, K., Kesselman, C.: CC++: A Declarative Concurrent Object Oriented Programming Notation, In Agha, G., Wegner, P., and Yonesawa (Eds.), Research Directions in Concurrent Object Oriented Programming, Cambridge, MA, MIT Press, pp. 218–313, 1993.
5. OMG: CORBA/IIOP 2.2 Specification, URL:http://www.omg.org (1998).
6. The Staff, Advanced Computing Laboratory, Los Alamos National Laboratory: Taming Complexity in High-Performance Computing. White Paper. Accessible from URL:http://www.acl.lanl.gov/software (November, 1998).

7. Reynders, J. et. al.: Pooma: A Framework for Scientific Simulation on Parallel Architectures, In: Wilson, G., Lu, P. (Eds.): Parallel Programming using C++, M.I.T. Press (1996) 553–594.
8. Miller, B., Callaghan, M., Cargille, J., Hollingsworth, J., Irvin, R., Karavanic, K., Kunchithapadam, K., Newhall, T.: The Paradyne Parallel Performance Measurement Tools, IEEE Computer. Vol. 28(11) (November, 1995).
9. Shende, S., Cuny, J., Hansen, L., Kundu, J., McLaughry, S., Wolf, O.: Event and State Based Debugging in TAU: A Prototype, Proc. of ACM SIGMETRICS Symp. on Parallel and Distributed Tools (May, 1996) 21–30.
10. Cuny, J., Dunn, R., Hackstadt, S., Harrop, C., Hersey, H., Malony, A., Toomey, D.: Building Domain-Specific Environments for Computational Science: A Case Study in Seismic Tomography, Intl. Jour. of Supercomputing Applications and High Performance Computing. Vol. 11 (March, 1997).
11. Shende, S.: Profiling and Tracing in Linux, In Proc. Second Extreme Linux Workshop #2, USENIX Annual Technical Conference (1999) 26–30.
12. Hackstadt, S., Harrop, C., Malony, A.: A Framework for Interacting with Distributed Programs and Data, In: Proc. of the Seventh Int'l Symp. on High Performance Distributed Computing 1998 (HPDC-7), IEEE (July, 1998).
13. Hackstadt, S., Malony, A.: DAQV: Distributed Array Query and Visualization Framework, Journal of Theoretical Computer Science, special issue on Parallel Computing Vol. 196, No. 1-2 (April, 1998) 289–317.
14. Shende, S., Malony, A. D., Hackstadt, S.: Dynamic Performance Callstack Sampling: Merging TAU and DAQV. In Kågström, B. and Dongarra, J. and Elmroth, E. and Waśniewski, J. (editors). Applied Parallel Computing, 4th International Workshop, PARA'98, Lecture Notes in Computer Science, No. 1541, Springer-Verlag, Berlin (June, 1998) 515-520.
15. Shende, S., Malony, A. D., Cuny, J., Lindlan, K., Beckman, P., Karmesin, S.: Portable Profiling and Tracing for Parallel, Scientific Applications using C++, Proc. of ACM SIGMETRICS Symp. on Parallel and Distributed Tools (Aug, 1998) 134-145.
16. Advanced Computing Laboratory (LANL): TAU Portable Profiling URL:http://www.acl.lanl.gov/tau. 1998.
17. Ludwig, T., Wismüller, R., Sunderam, V., and Bode, A.: OMIS – On-line Monitoring Interface Specification (Version 2.0), Vol. 9, LRR-TUM Research Report Series, Springer Verlag, Aachen, Germany, ISBN 3-8265-3035-7 (1997).
18. Wismüller, R., Trinitis, J., and Ludwig, T.: OCM – A Monitoring System for Interoperable Tools, Proceedings of the SIGMETRICS Symposium on Parallel and Distributed Tools (August, 1998) 1–9.
19. Ribler, R., Vetter, J., Simitci, H., and Reed, D.: Autopilot: Adaptive Control of Distributed Applications, Proceedings of the 7th IEEE International Symposium on High Performance Distributed Computing (July, 1998) 172–179.
20. Eisenhauer, G. and Schwan, K.: An Object-Based Infrastructure for Program Monitoring and Steering, Proceedings of the SIGMETRICS Symposium on Parallel and Distributed Tools, (August 1998) 10–20.

SharedOnRead Optimization in Parallel Object-Oriented Programming

Denis Caromel[1], Eric Noulard[2], and David Sagnol[1]

[1] University of Nice Sophia Antipolis, France
[2] Société ADULIS, France

Abstract. This paper presents how a specific technique (SharedOn-Read) can be used to improve performances of a distributed objects language. We present results for a network of workstations, providing comparisons between MPI and C++// implementations, both on homogeneous and heterogeneous platforms. The results could be applied to other models of distribution, such as RMI.

1 Introduction

This paper presents a mechanism for sharing objects when two processes (active objects [1] in our object-oriented framework) want to access the same object. This mechanism is implemented using the C++// [2,3] library. In C++//, active object cannot share passive ones, and there is a deep copy of all the parameters of a function call, between 2 active objects (the same behavior occurs in RMI or in Corba). This can be a bottleneck when active object are implemented using threads, on SMP machines.

This article proposes a solution to solve that problem: the SharedOnRead mechanism. A SharedOnRead object implements a well-known mechanism allowing to have both constant semantics and efficiency, whether objects are located in the same address space or not.

We test the SharedOnRead mechanism on basic linear algebra operations commonly used in iterative methods [4]. The key point for efficient parallel implementation of iterative methods are good performance of the distributed sparse matrix/vector product, and distributed dot product since these operations are at the heart of all basic Krylov algorithms [5]. In this context it is crucial to avoid unwanted copies in matrix operations. Explicitly dealt with in MPI-based parallel numerical applications, we want to achieve this in a more implicit and transparent way in an object model.

The next section gives a quick survey of the active object model we are using, presenting the main features of the C++// library. Section 3 introduces the SharedOnRead extension to the model, and Sect. 4 presents some performance results for matrix operation in linear algebra. Section 5 concludes the paper.

S. Matsuoka et al. (Eds.): ISCOPE'99, LNCS 1732, pp. 182–193, 1999.

2   The Active Object Model

The C++// language was designed and implemented with the aim of importing reusability into parallel and concurrent programming. It does not extend the language syntax, neither requires to modify the C++ compiler, since C++// is implemented as a library and a preprocessor (a Meta-Object Protocol [6]), and uses Nexus/Globus [7,8] library as the transport layer for communications. The C++// model adopts the following principles:

- heterogeneous model with both passive and active objects (actors, threads),
- active objects which act as sequential processes serving requests (i.e. method invocations) in a centralized and explicit manner by default,
- systematic asynchronous communications towards active objects,
- wait-by-necessity (automatic and transparent futures),
- automatic continuations (a transparent delegation mechanism),
- no shared passive objects (call-by-value between processes),
- centralized and explicit control by default,
- polymorphism between standard objects, active objects, and remote objects.

As shown in Fig. 1 (page 184), C++// programs are organized in subsystems: one subsystem contains *one* active object and *several* passive objects. Active objects are active by themselves, with their own thread of control, unlike passive objects that are normal C++ objects. An active object is the only one that can be referenced by objects in other subsystems; there are no shared passive objects. The absence of sharing is enforced by the programming model since all passive objects used as parameters are automatically transmitted by copy (deep copy of objects, alike serialization in Java RMI or other systems like ProActive [9]) when communication occurs towards an active object (another subsystem).

With respect to this paper, a crucial point is where active objects are created. A C++// programmer has several choices to determine the machine (or node) where a new active object will be created: (1) give the machine name, (2) provide an existing active object in order to use the same machine. We do not detail this mechanism here, see [2]. But central to the issue is that, in both cases, the programmer has two options to create the new active object:(a) in an existing address space, (b) in a new address space. In case (a), several active objects will be able to share the same address space — threads are used to implement active objects. Let us note that even when active objects in the same address space communicate, passive objects are still transmitted by copy. This potentially time and space consuming strategy is mandatory if we want the program semantics to be constant, whatever the mapping is. However, in some cases, sharing is actually possible, and copying large objects could be avoided. The SharedOnRead mechanism was defined to make possible such optimization, and to provide a general strategy that keeps the semantics constant when architecture and mapping vary.

3   The SharedOnRead Framework

The general idea of the strategy can be stated as follows: upon a communication between two subsystems that are within the same address space, instead of copying a SharedOnRead parameter, we just share it; there is no change to the copy semantics if the two subsystems are not in the same address space (cf. Fig. 1). The SharedOnRead is closely related to the *copy-on-write* mechanism that can be found in operating systems (Mach [10] or Orca [11,12] are using it). However, the strategy is slightly different in copy-on-write technique: you want to copy only when it is necessary instead, with the SharedOnRead mechanism, you want to share data on read operations when it is possible. The mechanism occurs at the user level and is just an optimization compared to the standard behavior. We do not have to deal with object coherency like in distributed shared memory since C++// does not provide distributed shared memory. Our main goal is to automatically optimize local computation that is explicitly dealt with in data-parallelism.

While this scheme is quite simple, a mechanism is needed in order to maintain the same semantics that occurs when the two subsystems are mapped in two different address spaces. We can notice that objects will be able to be shared by

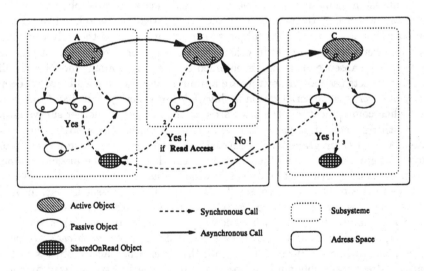

Fig. 1. SharedOnRead objects within several address spaces

several subsystems only as long as they are not modified by one of the subsystems. In order to be accurate, the strategy needs to make a distinction between *read* and *write* accesses, and also needs to know when a subsystem *forgets* a SharedOnRead object (the subsystem suppresses his reference to the SharedOnRead object).

The algorithm can be summarized as follows:

- When a SharedOnRead object is used as a communication parameter between two subsystems being in the same address space, the original object is not copied, but instead a new reference is memorized (a counter is incremented within that object).
- A *read* access is freely achieved (from both the owner's subsystem or another one).
- Upon a *write* access (from both the owner's subsystem or another one), if the counter value is more than 1, a copy of the object is made. The modification applies to the copy. The counter of the previously existing object is decremented; the counter of the copy is set to 1.
- Upon a *forget* operation, the counter is just decremented. When reaching zero, the object is automatically garbaged.

3.1 Programmer Interface and Implementation

As a design decision, we choose to give users the control over which objects should be SharedOnRead and which one should have the standard systematic copy behavior. A SharedOnRead object is an instance of a class that inherits directly or indirectly from the C++// SharedOnRead class (Fig. 2, class SharedOnRead). This class provides several member functions:

- *read_access* and *write_access*: both used to specify how the data members are accessed by a given method.
  - If *read_acess* is selected for a method, we ensure that, for this function, data members are never changed.
  - if *write_access* is selected for a method, we ensure that, for this function, data members can be changed.

  These two functions take a *mid_type* parameter which is unique for all the member functions in the program; the *mid* function[1] provides that unique identifier for a function name.
- The *access_declaration* function has to be redefined in order to specify for each public member function its read or write nature: write access is the default behavior. SharedOnRead user has to take care of that and check for each function in the class which ones are leaving the object in the same state and which ones are making modification to the object state.
- Finally, *forget* signals that we do not use the object anymore. This has to be dealt with explicitly because SharedOnRead object cannot be aware that they are not referenced anymore.

The implementation of the SharedOnRead class is based on the reification mechanism provided by C++//. Inheriting from the class *EC_Reflect* (cf. Fig. 2, class SharedOnRead), all the SharedOnRead member functions are reified [13]: the functions are not directly executed but go first in a specific *proxy* where all the

---

[1] this function is part of the C++// library

```
class SharedOnRead: public EC_Reflect {
public:
    virtual void read_access(mid_type);
    virtual void write_access(mid_type);
    virtual void access_declaration();
    virtual void forget();
};
class Block { // A Matrix block
public:
    virtual void reach(int** ai, int** aj, double** a);
    virtual void update(int**& ai, int**& aj, double**& a);
};
class SorBlock: public SharedOnRead, public Block {
public:
    SorBlock();
    virtual void access_declaration() {
        read_access(mid(reach));
        write_access(mid(update));
    }
};
```

Fig. 2. Programming a SharedOnRead block in a matrix example

necessary implementation is defined and achieved (update of the counter, copy when necessary, etc.). After this step, the *proxy* executes the function.

When a C++// programmer wants to get the SharedOnRead optimization, he just has to define a new class that inherits from an existing class and the SharedOnRead one, the only constraint being to redefine the *access_declaration* function. For instance, in the case of a *Block* class used to implement a matrix, a new class *SorBlock* can be defined, inheriting from both *Block* and SharedOnRead (Fig. 2). The function *access_declaration* specifies that *update* can make modification to the block, while *reach* cannot. This is the user responsibility since we cannot prevent a user from declaring a function with read access that actually modifies the object.

3.2 A Simple Example

The purpose of this section is to present a simple application using the SharedOnRead mechanism. The *Buffer* class (Fig. 3) has two virtual functions: one to get a particular element (getElem, the *Element* object can be any kind of standard C++ object) and one to modify an element (updateElem).

At the beginning of the program, this buffer is created by an active object (named *active1*) and is filled with different elements.

```
// active1 class
Buffer* buf1 = new Bufer(size); // Buffer creation
buf1->updateElem(2, e1);  // e1 is an Element
```

```
class Buffer: public SharedOnRead {
public:
    virtual Element* getElem(int i); // return element i
    virtual void updateElem(int i, Element* e); // update element i
    virtual void access_declaration() {
        read_access(mid(getElem));
        write_access(mid(updateElem));
    }
};
```

Fig. 3. A SharedOnRead buffer

At a moment of the calculation, another active object (named e.g. *active2*) wants to acquire a reference to this buffer object:

```
// active2 class
Buffer* buf2 = active1->getBuffer();
    // Get a direct pointer if we are in the
    // same address space
    // buf1 and buf2 point to the same object
Element* e = buf2->getElem(3); // get the third element
```

In non optimized C++// (without SharedOnRead), it would have acquired a deep copy of this buffer, but as *Buffer* inherits from SharedOnRead (and since *active1* and *active2* belong to the same address space) it acquires a direct (reified) pointer to the buffer. As long as the two active objects only use *getElem* functions, they can share the buffer. But if one makes an *updateElem*, it automatically acquires a copy before the function is executed. After that, *active1* and *active2* objects are working on two different buffers.

This example demonstrates that the main use of SharedOnRead objects is when we do not want to replicate big objects: several active objects can share the same SharedOnRead object. Even for small objects, the mechanism can be still interesting: it is better (smaller) to transmit just a reference to an object than to copy it. The choice should be guided by the use of the SharedOnRead object: because of reification overhead (a function call to a reified object is around 5 times the cost of a virtual call) it can be more interesting to achieve a copy and after that, to work on this copy. The choice depends on your primary goal, reduce memory usage or increase speed. Our system is able to generate traces to help a programmer to choose which is the best solution. This mechanism could also be used with the overlapping mechanism presented in [14]. If a small portion in a huge object is needed by an object in a different address space, the small portion could be send first as a SharedOnRead object and the rest would be sent later.

4   Benchmark Application: Parallel Linear Algebra

We will apply our SharedOnRead mechanism to the implementation of efficient matrix/matrix operations for (block) iterative methods. As defined in [5] we may focus on a reduced set of operations:

- dense SAXPY, $Y := \alpha X + \beta Y$ with $Y, X \in \mathcal{R}^{n \times p}, \alpha, \beta \in \mathcal{R}$.
- dense or sparse matrix product $Y := \alpha A \cdot X + \beta Y$ with $Y \in \mathcal{R}^{m \times p}$, $A \in \mathcal{R}^{m \times n}, X \in \mathcal{R}^{n \times p}, \alpha, \beta \in \mathcal{R}$

Parallel numerical linear algebra is concerned with data distribution of the matrix arguments in the above operations. In our case we will only consider a block distribution scheme of matrices which is widely used [15] and well suited for our applications. With these kind of distributions, we have to do dense matrix saxpy operation and dense or sparse matrix product.

These operations are implemented as a method of a class Matrix representing a block partitioned matrix. The methods are:

- Matrix::scal(double alpha, Matrix* B, double beta)
  This method performs the matrix saxpy operation $this = \beta \cdot this + \alpha \cdot B$.
- Matrix::axpy(double alpha, Matrix* A, Matrix* X, int mm)
  This method performs the matrix operation $this = \alpha \cdot A \times X + this$, and $this = \alpha \cdot A \times X$ if mm==1.

Here the distributed objects are the blocks of the matrix which may be called CSC since they are Compressed Sparse Column (potentially sparse) matrices.

4.1   From Sequential to Parallel Matrices in C++//

A sequential matrix contains a list of CSC objects, each of these objects holding a *Block*. This *Block* is responsible for allocating all the arrays representing the matrix. If we want to parallelize these classes using C++//, we only have to redefine the *Matrix* constructors. These constructors create the distributed CSC objects and associate (or not) the *SorBlock* depending if we want to use the SharedOnRead mechanism or not. A distributed CSC object (*CSC_ll* object) can be created just by inheriting from CSC and the C++// class *Process*. This class implements all the mechanisms we need to create active objects (remote creation, automatic marshalling, polymorphism between sequential and distributed objects, ...). This is the only part of the sequential code that have to be modified. All the functions presented above (scal, axpy, ...) come unchanged from the sequential classes.

Figure 4 presents the sequential version for a dense axpy function. Because of polymorphism compatibility, the two A and X variables can be *CSC_ll* objects. In this case, the *block()* function returns a *SorBlock* object (if SharedOnRead is used): we get access to the original object implementing the matrix block without copy. Since, it is sufficient to access *bl1* and *bl2* in read mode ; if they are located in the same address space, we use them directly, otherwise we get a local copy. On the other hand, the *mine* variable, which represents the local *Block* of the *CSC* object, has to be modify so it must be "declared" in write mode.

```
void CSC::axpy(double alpha, CSC* A, CSC* X, double beta) {
    Block* bl1 = A->block();
    Block* bl2 = X->block();
    Block* mine = block();
    bl1->reach(&tia, &tja, &ta);
    bl2->reach(&tix, &tjx, &tx);
    mine->update(&tiy,&tjy,&ty);
    ...
}
```

Fig. 4. Sequential dense CSC_Block product

4.2 MPI Version

The MPI implementation requires redefinition of all the functions because we do not perform the same operations depending on the processor we are executing on: MPI is very intrusive. This means that you have to add many MPI calls to the sequential code in order to obtain the parallel version. Moreover, this makes it difficult for the application developer to go back and forth between sequential and parallel versions. Further, since MPI is a message passing library, parallelism is explicit, and the programmer must directly deal with distribution and communication. Another advantage of the C++//version is that we do not require use of a temporary matrix for the reduction step. In the examples in the next section, this temporary matrix can take up to 7MB.

Even if C++// is not a pure data parallel language, we choose to compare our results with MPI [16]. MPI is a well known and efficient standard library to program parallel applications.

4.3 Performance

The following tests were performed on a network of 4 Solaris Ultra 1 with 128MB of memory and a 10Mb Ethernet link. The MPI tests use the Lam library.[2]

Since the platform is message passing, several active objects (the CSC or the matrix blocs) are mapped within the same address space on each workstation using a standard distribution algorithm. The SharedOnRead optimization works when a computation occurs within the same address space. As demonstrated below this is sufficient to achieve consequent speedup. If we were on an SMP architecture, then the benefits would be even greater since there would be opportunity for sharing between all the matrix blocks.

---

[2] http://www.mpi.nd.edu/lam/

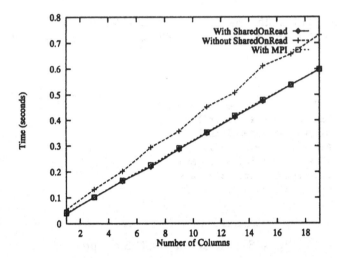

Fig. 5. SAXPY with 4 computers

Figure 5 presents our result for a scal (cf. 4) calculation. Matrices used during these tests were rectangular matrices with 90449 rows and a variable number of columns. The use of SharedOnRead objects demonstrates a speed-up between 20 and 25% compared to the non optimized C++// version. When compared with the MPI version, we cannot distinguish any difference between C++// with SharedOnRead and MPI. One important point to notice is the fact that the non optimized C++// version (without SharedOnRead objects) presents more and more overhead when the matrix size is growing. The main reason being that this version requires many communications between the different active objects that are located in different processes. For the SharedOnRead version, the two blocks (two active objects) are mapped within the same process, so we avoid communication and copy.

Figure 6 presents results for a dense matrix product. Again, the non optimized version is between 20 and 30% slower than MPI. Between the C++// version and the MPI one, the overhead is constant. The SharedOnRead version and the MPI one do not present exactly the same results in this test because this operation requires communication during the reduction phase. The C++// solution sums the local matrix in a sequential way because it reuses the sequential code. Figure 7 presents the speed-up obtained with 4 computers for the 3 different configurations. All calculations on dense matrices are perfectly scalable; for 4 machines, the speed-up is around 3.9. We can observe that the overhead of the non optimized C++// version is constant whatever the number of computers we use.

Finally, Fig. 8 presents results for a sparse matrix product. A first point to notice in such a case is that we had to rewrite the *add* function of the matrix. The reduction being critical in this benchmark, it was important to achieve it in

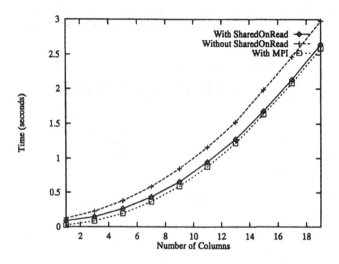

Fig. 6. Dense matrix product with 4 computers

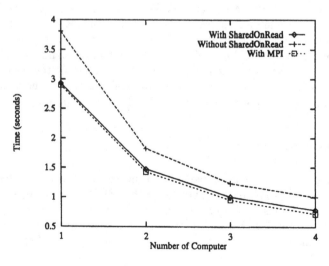

Fig. 7. Dense matrix product with (90449x10) Matrix

parallel. The second important aspect of this benchmark deals with the platform architecture. All the previous tests were made on homogeneous computers: the same CPU, at the same frequency, with the same amount of memory. This last test was performed on an heterogeneous platform. Two Ultra 1 were replaced

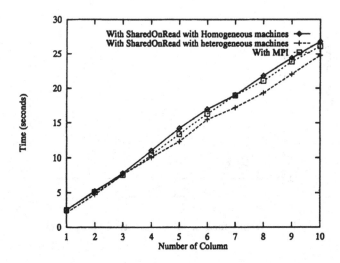

Fig. 8. Sparse matrix product with 4 computers

with two Ultra 2 with 192 MB of memory. Within this new architecture, the MPI program has the same performance as in the homogeneous test. In the C++// case, the benchmark demonstrates that the object-oriented version is actually more efficient than the MPI one. While MPI is subject to synchronization barriers, the C++// version automatically takes advantage of the asynchronous model. In that case, the reduction of the two local matrices of the fastest computers can start even if the computation on the slowest computers is not finished.

5 Conclusion

This paper presented how a SharedOnRead mechanism can help a distributed object-oriented language to be competitive with MPI. In some cases, we can even overcome MPI thanks to C++// asynchronism and wait-by-necessity mechanisms. When comparing the MPI implementation with the C++// one, it also appears that the later allows us to avoid writing large amount of code to obtain a parallel version.

This technique of passive object sharing could be implemented in any object-oriented language with distributed objects that do not have a mechanism to share passive object upon copy semantics (RMI in Java, CORBA, etc.). It allows to bring together object-orientedness, simplicity, reusability and performances.

References

1. Gul Agha. *A Model of Concurrent Computation in Distributed Systems.* MIT Press, 1986.
2. D. Caromel, F. Belloncle, and Y. Roudier. The C++// Language. In *Parallel Programming Using C++*, pages 257–296. MIT Press, 1996.
3. Caromel D., Dzwig P., Kauffman R., Liddell H., McEwan A., Mussi P., Poole J., Rigg M., and Winder R. Ec++ – europa parallel c++ : A draft definition. In *Proceedings of High-Performance Computing and Networking (HPCN'96)*, volume 1067, pages 848–857. Springer, Lectures Notes in Computer Science (LNCS), 1996.
4. Yousef Saad. *Iterative Methods for Sparse Linear Systems.* PWS Publishing Company, New York, 1996.
5. Eric Noulard, Nahid Emad, and Laurence Flandrin. Calcul numérique parallèle et technologies objet. Technical Report Rapport PRISM 1998/003, ADULIS/PRiSM, Juillet 1997. Révision du 30/01/98.
6. G. Kiczales, J. des Rivières, and D.G. Bobrow. *The Art of the Metaobject Protocol.* MIT Press, 1991.
7. I. Foster, C. Kesselman, and S. Tuecke. The Nexus approach to integrating multithreading and communication. *J. Par. Distr. Comp.*, 37:70–82, 1996.
8. Ian Foster and Carl Kesselman. Globus: A metacomputing infrastructure toolkit. *International Journal of Supercomputer Applications*, 11(2):115–128, 1997.
9. D. Caromel, W. Klauser, and J. Vayssiere. Towards Seamless Computing and Metacomputing in Java. *Concurrency Practice and Experience*, 1998.
10. R. Rashid, R. Baron, A. Forin, D. Golub, M. Jones, D. Orr, and R. Sanzi. Mach: a foundation for open systems (operating systems). In *Workstation Operating Systems: Proc. 2nd Workshop on Workstation Operating Systems (WWOS-II), September 27–29, 1989*, pages 109–113.
11. H. E. Bal, M. F. Kaashoek, A. S. Tanenbaum, and J. Jansen. Replication techniques for speeding up parallel applications on distributed systems. *Concurrency Practice & Experience*, 4(5):337–355, August 1992.
12. S. B. Hassen and H. Bal. Integrating task and data parallelism using shared objects. *Proc. FCRC '96: Proc. 1996 Int. Conf. Supercomputing*, pages 317–324.
13. D. Caromel, A. McEwan, J. Nolte, J. Poole, Y. Roudier, D. Sagnol, J.-M. Challier, P. Dzwig, R. Kaufman, H. Liddell, P. Mussi, D. Parkinson, M. Rigg, G. Roberts, and R. Winder. Europa parallel c++, Septembre 1997. The EUROPA Working Group on Parallel C++, Final report, HPCN Esprit Contract No 9502.
14. Françoise Baude, Denis Caromel, Nathalie Furmento, and David Sagnol. Overlapping Communication with Computation in Dis tributed Object Systems. In *Proceedings of the 7th International Conference - HPCN'99)*, volume 1593 of *Lecture Notes in Computer Science*, pages 744–753, 1999.
15. J. Choi, J. Dongarra, S. Ostrouchov, A. Petitet, D. Walker, and R.C. Whaley. A proposal for a set of parallel basic linear algebra subprograms. Technical Report Lapack Working Note 100, May 1995.
16. Message Passing Interface Forum. MPI: A message passing interface standard. Technical Report CS-94-230, University of Tennessee, Knoxville, TN, March 1994.

Branch and Bound Based Load Balancing for Parallel Applications

Shobana Radhakrishnan, Robert K. Brunner, and Laxmikant V. Kalé

University of Illinois at Urbana-Champaign, Urbana, IL, USA

Abstract. Many parallel applications are highly dynamic in nature. In some, computation and communication patterns change gradually during the run; in others those characteristics change abruptly. Such dynamic applications require an adaptive load balancing strategy. We are exploring an adaptive approach based on multi-partition object-based decomposition, supported by object migration. For many applications, relatively infrequent load balancing is needed. In these cases it becomes economical to spend considerable computation time toward arriving at a nearly optimal mapping of objects to processors. We present an optimal-seeking branch and bound based strategy that finds nearly optimal solutions to such load balancing problems quickly, and can continuously improve such solutions as time permits.

1 Introduction

Development of efficient parallel applications becomes difficult when they are either irregular or dynamic or both. In an irregular application, the computational costs of its subcomponents cannot be predicted accurately. Other applications are dynamic, with the computational costs of their subcomponents changing over time. In either case, performance problems manifest themselves in the form of load imbalances. Although such imbalances are typically small and tolerable while running applications on a small number of processors, they often become major performance drains on systems with a large number of processors.

We have been exploring a solution to this problem that involves breaking the problem into a large number of chunks, such that the total number of chunks is significantly larger than the number of available processors. In fact, the size of a chunk can be decided independently of the number of processors, by using the criterion of keeping the communication overhead within a pre-specified bound. A system that supports data driven objects, (*e.g.* Charm++ [1]) is used to implement each chunk as an independent object. Thus, these objects send messages to other *objects*, in contrast to an MPI program (for example), which directs messages to specific processors. As a result, the runtime system is free to move the objects from one processor to another, without disturbing the application. Charm++ supports such migration of objects with automatic and optimized forwarding of messages. With these prerequisites (multi-chunk object-based decomposition, and support for object migration), all that one needs is a strategy to decide when and where to move objects.

S. Matsuoka et al. (Eds.): ISCOPE'99, LNCS 1732, pp. 194–199, 1999.

Even in irregular and dynamic programs, one can find a basis for predicting future performance. Just as in sequential programs one can rely on the principle of locality, in a parallel program one can utilize the principle of "temporal persistence of computation and communication patterns". In irregular computations, each subcomponent's computation time may be unpredictable *a priori*, but once the program starts executing, each component will persist in its behavior over the iterations of the program. In dynamic applications, the behavior of a component changes, but even here, either the behavior changes slowly over time, or abruptly but infrequently (as in adaptive refinement strategies). In either case, it is a reasonable heuristic to assume such a persistence of behavior, over some horizon in the future. This is not unlike the idea of using caches based on the principle of locality and working sets. Although the program may jump out of its working set from time to time, the caching technique, which assumes that the data referenced in the recent past will continue to be referenced, still pays large performance dividends.

Based on the above performance prediction principle, we have developed an adaptive load balancing framework. It provides automatic measurement of computation times and automatic tracing of communication events of a parallel object program. A load balancing strategy can obtain the necessary performance data from this framework, and decide to migrate some objects to new processors.

Within the context of this framework, we are engaged in developing a suite of load balancing strategies, and applying them in a variety of applications. Different classes of applications require different load balancing strategies. In a significant class of applications, focused on in this paper, only periodic, and *infrequent* rebalancing is necessary.

Our experience with molecular dynamics [2] for biophysical simulations shows, for example, that the load balance stays relatively stable over several hours as the atoms slowly migrate over domain boundaries. In such a situation, spending as much as a few minutes on deciding a new mapping is not that expensive. However, the problem of optimum mapping is NP-hard. So, even with minutes of time on a parallel machine it typically will not be possible to find the provably optimal solution. One thus appears to be stuck between the bimodal choice of, a low-cost, low-quality heuristic method, or an unrealistic, optimum-finding algorithm. This paper presents a branch and bound based strategy that fills in the middle ground: depending on the available computation time, it can produce a continuum of solutions from the simple heuristic ones to provably optimal ones.

2   The Object Model

This section describes how our algorithm approaches the load balancing problem, by modeling parallel applications as collections of computation objects which communicate among themselves. Communication costs between objects are modeled based on the characteristics of the particular machine, and objects on the same processor are assumed to exchange data for free. Furthermore, the load balancer has the freedom to reassign these objects to any processors to optimize program performance.

The objects that are to be balanced are represented as a network of communicating entities in the form of a directed graph. Graph-based models have been used earlier for the task allocation problem (*e.g.* [3]). Also, Metis [4] provides a graph based partitioning scheme that is meant for partitioning large, million-element unstructured meshes. The vertices in the graph represent the computation cost of the objects to be balanced and each edge represents communication, parameterized by the pair <*number of messages, total bytes sent*>. If the sending and receiving objects are assigned to different processors, the processors are charged:

$$T_{send} = \alpha_{send} \cdot N_{messages} + \beta_{send} \cdot N_{bytes}$$
$$T_{receive} = \alpha_{receive} \cdot N_{messages} + \beta_{receive} \cdot N_{bytes}$$

In addition to migratable objects and communication patterns, our object model also includes the following features:

1. **Non-migratable Objects:** Non-migratable objects are objects which must remain on particular processors throughout their lifetime. Load balancers should still consider their computation and communication cost as background load, but do not have the freedom to move them.
2. **Proxy Communication:** This refers to multicast communication where several objects require data from one particular object. Should the receiving objects all be placed on the same processor, a single message may supply the data to all of the receivers. We model this by adding an attribute, the *proxy_id*, for each message arc. While calculating the communication cost resulting from the assignment of an object to a processor, we ignore the cost of an incoming multicast communication arc if another recipient of the same multicast has already been assigned to this processor.

3   Branch and Bound Algorithm

Branch and bound algorithms are a good choice for load balancers, because they exhibit the property that they can produce an optimal solution if given enough time, but produce "good" sub-optimal solutions if stopped prematurely. To provide the flexible tradeoff between decision time and solution-quality, we limit the load balancing algorithm to a caller-specified time limit. Although this usually does not let the algorithm pursue all possible states, our optimized algorithm still gives the solution quite close to optimal as compared to the other algorithms we have implemented.

Our branch and bound load balancer follows the design of common branch and bound algorithms, with the addition of a few optimizations particular to the load-balancing problem.

- **Sorting objects before assignment:** The objects are ordered in decreasing sequence of their computation costs for assignment. Thus, more expensive objects are assigned at higher levels of the search tree.
- **Search ordering:** At each level of the search tree, the child that assigns the new object to the least loaded processor is considered first.
- **Greedy Initial Estimate:** States are pruned based not on the first state evaluated. Instead, a quickly-obtained greedy estimate is used as the initial lower bound which results in more states being pruned early.
- **Symmetry:** If all the processors have identical communication and computation capacities, then any processor with no assigned objects is equivalent to another such processor. This reduces the branching factor of the tree at the top levels.
- **Future-Cost Estimates:** Instead of just using the costs of states previously assigned to obtain the current lower bound, we compute an optimistic estimate of the cost of assigning the remaining states to obtain a more accurate lower bound, which allows the search to prune more states.

As suggested by Wah and Yu [5], one could narrow the search space by aiming for a solution guaranteed to be within a small percentage (say two percent) of the optimal. This is accomplished by comparing the lower bound to $0.98 \times$ *upper_bound* in the pruning step. In the context of our strategy, which uses a fixed time limit, such a narrowing may seem to be even more beneficial, as it allows the search to "sample" a larger portion of the search space. However, in almost all the runs we conducted, with using 1, 2 and 4 percent tolerance, we found no improvement in solution quality within fixed time.

4   Performance Results

In this section, we compare the branch and bound load balancer with four other algorithms. These algorithms include:

1. **Greedy:** This algorithm uses a greedy heuristic without performing the branch and bound search.
2. **Random:** Objects are randomly distributed among the processors.
3. **Greedy-Refine:** The greedy algorithm is run to obtain an initial distribution, and then a refinement procedure is applied. This refinement procedure looks at each processor with a load above the average by a certain threshold, and moves objects from them to under-loaded processors, until no further movement is possible.
4. **Random-Refine:** The refinement procedure is applied to the solution found with the random algorithm.

All of these algorithm (except Random) consider the processor overhead of communication in the assignment process, in accordance with the cost model in Sect. 2.

Table 1 shows the results obtained when runs were made of the sequential implementation of the branch and bound strategy using a recursive method

Table 1. Efficiency

Case #	Procs.	Comm. Cost	Greedy	Greedy-Refine	Random	Random-Refine	Branch & Bound
1	9	0	99.7	99.7	69.1	69.1	99.8
2	20	0	98.4	98.4	57.5	57.5	99.4
3	9	120	51.4	55.6	58.5	68.6	81.0
4	20	120	28.8	31.7	50.6	67.7	78.4
5	9	250	34.4	37.0	48.4	55.9	64.4
6	20	250	26.3	28.5	41.2	44.7	60.1
7	9	300	37.1	40.9	46.0	50.9	60.3
8	20	300	26.7	30.0	39.1	42.1	56.2
9	9	400	44.2	52.2	41.8	50.5	54.6
10	20	400	21.2	24.0	35.4	36.9	49.6
11	9	500	26.9	28.9	38.4	46.4	49.5
12	20	500	27.4	30.0	32.3	42.3	43.7
13	9	600	29.9	34.7	35.4	41.7	44.3
14	20	600	13.6	14.2	29.6	38.0	39.5
15	9	700	20.9	22.2	32.9	38.4	41.1

for various cases. In all cases, the same object graph is used, with 100 objects and randomly generated computation cost and communication volumes. The *efficiency* is calculated as $T_{sequential}/(P \cdot T_{parallel})$, where P is the number of processors, and $T_{parallel}$ is computed by taking communication into account. We observe that, even when run for limited time (so that the search tree is not exhaustively searched), the branch and bound strategy gives the most efficient solution among the algorithms implemented.

From these results, we observe that the efficiency of the solution for each algorithm decreases as the communication overhead increases. This occurs because the optimal efficiency itself goes down with increase in the communication overhead.

We also monitored the quality of solution as a function of time spent by the load balancer. As expected, the quality increases with more search, but at some time it converges on an optimum value. It can be verified from small problem instances, that further search time spent on proving the near-optimality of the solution quickly exceeds the time savings resulting from the slightly improved load balance. This result is consistent with observations in the operations research community regarding hard search problems.

We observed that applying the refinement algorithm does not greatly increase the time spent by any of the load balancing algorithms, but produces a much better solution in many cases. For example, in most cases the refine applied to Greedy takes about 1 second more, and results in about 10 percent efficiency improvement. For Random, refinement requires proportionally more time, but the resulting efficiency is improved even more dramatically.

Often load balancing strategies concentrate on balancing computation costs alone. To understand the effects of ignoring communication, we evaluated the performance of the algorithms after modifying them to ignore communication costs. For instance, we found that the modified Random-Refine strategy led to an efficiency of 33 percent, compared to 39 percent obtained with original strategy.

5   Summary and Planned Work

In this paper, we presented a branch and bound based strategy that uses this data to generate a near-optimal mapping of objects to processors. This strategy is a component of our object based load balancing infrastructure to effectively parallelize irregular and dynamic applications. The framework instruments parallel programs consisting of intercommunicating parallel objects, and collects performance and communication data. A useful property of the branch and bound strategy is that it is tunable: it has the ability to use the available time to produce increasingly better mappings. Also, the object communication costs are fully modeled. Intelligent greedy strategies were also developed, and are seen to be quite effective. The new strategy performs satisfactorily, irrespective of the communication to computation ratio.

The branch and bound algorithm itself is suitable for execution in parallel; indeed we have developed such a parallel variant. Using this, we plan to conduct extensive performance studies. In particular, we plan to perform further studies using various parallel machines and applications, rather than just the simulation model described in this paper.

Due to space limitations, this paper does not include a survey and comparison with the extensive load balancing literature. We only note that many strategies described in the literature are either not oriented toward an object-graph model or do not present a tunable strategy.

References

1. L. V. Kale and Sanjeev Krishnan. Charm++: Parallel Programming with Message-Driven Objects. In Gregory V. Wilson and Paul Lu, editors, *Parallel Programming using C++*, pages 175–213. MIT Press, 1996.
2. Laxmikant Kalé, Robert Skeel, Milind Bhandarkar, Robert Brunner, Attila Gursoy, Neal Krawetz, James Phillips, Aritomo Shinozaki, Krishnan Varadarajan, and Klaus Schulten. NAMD2: Greater scalability for parallel molecular dynamics. *Journal of Computational Physics*, 1998. In press.
3. P. M. A. Sloot A. Schoneveld, J. F. de Ronde. Preserving locality for optimal parallelism in task allocation. In *HPCN*, pages 565–574, 1997.
4. George Karypis and Vipin Kumar. Parallel multilevel k-way partitioning scheme for irregular graphs. In *Proc. Supercomputing '96*, Pittsburg, PA, November 1996.
5. B. W. Wah and C. F. Yu. Stochastic modeling of branch-and-bound algorithms with best-first search. *IEEE TSE*, 11:922–934, 1985.
6. Chengzhong Xu and Francis C. M. Lau. *Load Balancing In Parallel Computers Theory and Practice*. Kluwer Academic Publishers, 1997.

Author Contacts

Federico Bassetti, Los Alamos National Laboratory, Los Alamos, NM 87545, USA; fede@lanl.gov.

David L. Brown, Lawrence Livermore National Laboratory, Livermore, CA 94551, USA; dlb@llnl.gov.

David L. Bruhwiler, Tech-X Corporation, 1280 28th Street, Suite 2, Boulder, CO 80303, USA; bruhwile@txcorp.com.

Robert K. Brunner, Department of Computer Science, University of Illinois at Urbana-Champaign, Urbana, IL 61801, USA; rbrunner@cs.uiuc.edu.

Denis Caromel, University of Nice Sophia Antipolis, INRIA, 2004 Route des Lucioles, B.P. 93, 06902 Valbonne Cedex, France; Denis.Caromel@sophia.inria.fr.

John R. Cary, Tech-X Corporation, 1280 28th Street, Suite 2, Boulder, CO 80303, USA; cary@txcorp.com.

Amar Chaudhary, Center for New Music and Audio Technologies, University of California, Berkeley, CA 94709, USA; amar@cnmat.berkeley.edu.

Kei Davis, Los Alamos National Laboratory, Los Alamos, NM 87545, USA; kei@lanl.gov.

Nathan Dykman, Center for Applied Scientific Computing, Lawrence Livermore National Laboratory, Livermore, CA 94551, USA; dykman1@llnl.gov.

Noah Elliott, Center for Applied Scientific Computing, Lawrence Livermore National Laboratory, Livermore, CA 94551, USA; elliott22@llnl.gov.

Adrian Freed, Center for New Music and Audio Technologies, University of California, Berkeley, CA 94709, USA; adrian@cnmat.berkeley.edu.

Nobuhisa Fujinami, Sony Computer Science Laboratories Inc., Takanawa Muse Building, 3-14-13 Higashi-gotanda, Shinagawa-ku, Tokyo 141-0022, Japan; fnami@csl.sony.co.jp.

Jens Gerlach, Real World Computing Partnership, GMD-FIRST, Kekulé Straße 7, 12489 Berlin, Germany; jens@first.gmd.de.

Dhrubajyoti Goswami, Department of Electrical and Computer Engineering, University of Waterloo, Waterloo, Ontario N2L-3G1, Canada; goswami@etude.uwaterloo.ca.

Edwin Günthner, Computer Science Department, University of Karlsruhe, Am Fasanengartenn 5, 76128 Karlsruhe, Germany; edwin.guenthner@gmx.de.

Youn-Hee Han, Department of Computer Science and Engineering, Korea University, 5-1 Anam-dong, Seongbuk-ku, Seoul 136-701, Republic of Korea; yhhan@disys.dorea.ac.kr.

William D. Henshaw, Lawrence Livermore National Laboratory, Livermore, CA 94551, USA; henshaw1@llnl.gov.

Chong-Sun Hwang, Department of Computer Science and Engineering, Korea University, 5-1 Anam-dong, Seongbuk-ku, Seoul 136-701, Republic of Korea; hwang@disys.dorea.ac.kr.

Young-Sik Jeong, Department of Computer Science and Engineering, Korea University, 5-1 Anam-dong, Seongbuk-ku, Seoul 136-701, Republic of Korea; ysjeong@disys.dorea.ac.kr.

S. Matsuoka et al. (Eds.): ISCOPE'99, LNCS 1732, pp. 201–203, 1999.
© Springer-Verlag Berlin Heidelberg 1999

Xiangmin Jiao, Department of Computer Science, University of Illinois at Urbana-Champaign, Urbana, IL 61801, USA; jiao@uiuc.edu.

Laxmikant V. Kal'e, Department of Computer Science, University of Illinois at Urbana-Champaign, Urbana, IL 61801, USA; kale@cs.uiuc.edu.

Scott Kohn, Center for Applied Scientific Computing, Lawrence Livermore National Laboratory, Livermore, CA 94551, USA; kohn1@llnl.gov.

Lie-Quan Lee, Laboratory for Scientific Computing, Department of Computer Science and Engineering, University of Notre Dame, Notre Dame, IN 46556, USA; llee1@lsc.nd.edu.

Xiang-Yang Li, Department of Computer Science, University of Illinois at Urbana-Champaign, Urbana, IL 61801, USA; xli2@uiuc.edu.

Edward A. Luke, NSF Engineering Research Center, Mississippi State University, Box 9627, Mississippi State, MS 39762, USA; lush@erc.msstate.edu.

Andrew Lumsdaine, Laboratory for Scientific Computing, Department of Computer Science and Engineering, University of Notre Dame, Notre Dame, IN 46556, USA; lums@lsc.nd.edu.

Xiaosong Ma, Department of Computer Science, University of Illinois at Urbana-Champaign, Urbana, IL 61801, USA; xma1@uiuc.edu.

A. Malony, Computational Science Institute, Department of Computer and Information Science, University of Oregon, Eugene, OR 97403, USA; malony@cs.uoregon.edu.

Madhav Marathe, Los Alamos National Laboratory, Los Alamos, NM 87545, USA; marathe@lanl.gov.

Serge Miguet, Laboratoire ERIC, Université Lumière Lyon 2, Bat. L, 5 Av Pierre Mendès-France, 69676 Bron Cedex, France; miguet@univ-lyon2.fr.

Eric Noulard, Société ADULIS, 3 Rue René Cassin, F-91724 Massy Cedex, France; E.Noulard@adulis.fr.

Chan Yeol Park, Department of Computer Science and Engineering, Korea University, 5-1 Anam-dong, Seongbuk-ku, Seoul 136-701, Republic of Korea; chan@disys.dorea.ac.kr.

Bobby Philip, Lawrence Livermore National Laboratory, Livermore, CA 94551, USA; philip1@llnl.gov.

Michael Philippsen, Computer Science Department, University of Karlsruhe, Am Fasanengarten 5, 76128 Karlsruhe, Germany; phlipp@ira.uka.de.

Bruno R. Preiss, Department of Electrical and Computer Engineering, University of Waterloo, Waterloo, Ontario N2L-3G1, Canada; brpreiss@despot.uwaterloo.ca.

Dan Quinlan, Lawrence Livermore National Laboratory, Livermore, CA 94551, USA; dquinlan@llnl.gov.

Shobana Radhakrishnan, Department of Computer Science, University of Illinois at Urbana-Champaign, Urbana, IL 61801, USA; rdhkrshn@cs.uiuc.edu.

Dhananjai Madhava Rao, Computer Architecture Design Laboratory, Dept. of ECECS, P.O. Box 210030, University of Cincinnati, Cincinnati, OH 45221-0030, USA; dmadhava@ececs.uc.edu.

David Sagnol, University of Nice Sophia Antipolis, INRIA, 2004 Route des Lucioles, B.P. 93, 06902 Valbonne Cedex, France;David.Sagnol@sophia.inria.fr.

David Sarrut, Laboratoire ERIC, Université Lumière Lyon 2, Bat. L, 5 Av Pierre Mendès-France, 69676 Bron Cedex, France; dsarrut@univ-lyon2.fr.

Mitsuhisa Sato, Real World Computing Partnership, GMD-FIRST, Kekulé Straße 7, 12489 Berlin, Germany; msato@trc.rwcp.or.jp.

Svetlana G. Shasharina, Tech-X Corporation, 1280 28th Street, Suite 2, Boulder, CO 80303, USA; sveta@txcorp.com.

T. Sheehan, Computational Science Institute, Department of Computer and Information Science, University of Oregon, Eugene, OR 97403, USA; sheehan@cs.uoregon.edu.

S. Shende, Computational Science Institute, Department of Computer and Information Science, University of Oregon, Eugene, OR 97403, USA; sameer@cs.uoregon.edu.

Jeremy G. Siek, Laboratory for Scientific Computing, Department of Computer Science and Engineering, University of Notre Dame, Notre Dame, IN 46556, USA; jsiek@lsc.nd.edu.

Ajit Singh, Department of Electrical and Computer Engineering, University of Waterloo, Waterloo, Ontario N2L-3G1, Canada; asingh@etude.uwaterloo.ca.

Brent Smolinski, Center for Applied Scientific Computing, Lawrence Livermore National Laboratory, Livermore, CA 94551, USA; smolinski1@llnl.gov.

Nenad Stankovic, Department of Computing, Macquarie University, Sydney, NSW 2109, Australia; nstankov@ics.mq.edu.au.

Matthias Troyer, Theoretische Physik, ETH Zürich, CH-8093 Zürich, Switzerland; troyer@itp.phys.ethz.ch.

David Wessel, Center for New Music and Audio Technologies, University of California, Berkeley, CA 94709, USA; wessel@cnmat.berkeley.edu.

Philip A. Wilsey, Computer Architecture Design Laboratory, Dept. of ECECS, P.O. Box 210030, University of Cincinnati, Cincinnati, OH 45221-0030, USA; philip.wilsey@uc.edu.

Kang Zhang, Department of Computing, Macquarie University, Sydney, NSW 2109, Australia, kang@ics.mq.edu.au.

Author Index

Lecture Notes in Computer Science

For information about Vols. 1–1663
please contact your bookseller or Springer-Verlag

Vol. 1698: M. Felici, K. Kanoun, A. Pasquini (Eds.), Computer Safety, Reliability and Security. Proceedings, 1999. XVIII, 482 pages. 1999.

Vol. 1699: S. Albayrak (Ed.), Intelligent Agents for Telecommunication Applications. Proceedings, 1999. IX, 191 pages. 1999. (Subseries LNAI).

Vol. 1700: R. Stadler, B. Stiller (Eds.), Active Technologies for Network and Service Management. Proceedings, 1999. XII, 299 pages. 1999.

Vol. 1701: W. Burgard, T. Christaller, A.B. Cremers (Eds.), KI-99: Advances in Artificial Intelligence. Proceedings, 1999. XI, 311 pages. 1999. (Subseries LNAI).

Vol. 1702: G. Nadathur (Ed.), Principles and Practice of Declarative Programming. Proceedings, 1999. X, 434 pages. 1999.

Vol. 1703: L. Pierre, T. Kropf (Eds.), Correct Hardware Design and Verification Methods. Proceedings, 1999. XI, 366 pages. 1999.

Vol. 1704: Jan M. Żytkow, J. Rauch (Eds.), Principles of Data Mining and Knowledge Discovery. Proceedings, 1999. XIV, 593 pages. 1999. (Subseries LNAI).

Vol. 1705: H. Ganzinger, D. McAllester, A. Voronkov (Eds.), Logic for Programming and Automated Reasoning. Proceedings, 1999. XII, 397 pages. 1999. (Subseries LNAI).

Vol. 1706: J. Hatcliff, T. Æ. Mogensen, P. Thiemann (Eds.), Partial Evaluation – Practice and Theory. 1998. IX, 433 pages. 1999.

Vol. 1707: H.-W. Gellersen (Ed.), Handheld and Ubiquitous Computing. Proceedings, 1999. XII, 390 pages. 1999.

Vol. 1708: J.M. Wing, J. Woodcock, J. Davies (Eds.), FM'99 – Formal Methods. Proceedings Vol. I, 1999. XVIII, 937 pages. 1999.

Vol. 1709: J.M. Wing, J. Woodcock, J. Davies (Eds.), FM'99 – Formal Methods. Proceedings Vol. II, 1999. XVIII, 937 pages. 1999.

Vol. 1710: E.-R. Olderog, B. Steffen (Eds.), Correct System Design. XIV, 417 pages. 1999.

Vol. 1711: N. Zhong, A. Skowron, S. Ohsuga (Eds.), New Directions in Rough Sets, Data Mining, and Granular-Soft Computing. Proceedings, 1999. XIV, 558 pages. 1999. (Subseries LNAI).

Vol. 1712: H. Boley, A Tight, Practical Integration of Relations and Functions. XI, 169 pages. 1999. (Subseries LNAI).

Vol. 1713: J. Jaffar (Ed.), Principles and Practice of Constraint Programming – CP'99. Proceedings, 1999. XII, 493 pages. 1999.

Vol. 1714: M.T. Pazienza (Eds.), Information Extraction. IX, 165 pages. 1999. (Subseries LNAI).

Vol. 1715: P. Perner, M. Petrou (Eds.), Machine Learning and Data Mining in Pattern Recognition. Proceedings, 1999. VIII, 217 pages. 1999. (Subseries LNAI).

Vol. 1716: K.Y. Lam, E. Okamoto, C. Xing (Eds.), Advances in Cryptology – ASIACRYPT'99. Proceedings, 1999. XI, 414 pages. 1999.

Vol. 1717: Ç. K. Koç, C. Paar (Eds.), Cryptographic Hardware and Embedded Systems. Proceedings, 1999. XI, 353 pages. 1999.

Vol. 1718: M. Diaz, P. Owezarski, P. Sénac (Eds.), Interactive Distributed Multimedia Systems and Telecommunication Services. Proceedings, 1999. XI, 386 pages. 1999.

Vol. 1719: M. Fossorier, H. Imai, S. Lin, A. Poli (Eds.), Applied Algebra, Algebraic Algorithms and Error-Correcting Codes. Proceedings, 1999. XIII, 510 pages. 1999.

Vol. 1720: O. Watanabe, T. Yokomori (Eds.), Algorithmic Learning Theory. Proceedings, 1999. XI, 365 pages. 1999. (Subseries LNAI).

Vol. 1721: S. Arikawa, K. Furukawa (Eds.), Discovery Science. Proceedings, 1999. XI, 374 pages. 1999. (Subseries LNAI).

Vol. 1722: A. Middeldorp, T. Sato (Eds.), Functional and Logic Programming. Proceedings, 1999. X, 369 pages. 1999.

Vol. 1723: R. France, B. Rumpe (Eds.), UML'99 – The Unified Modeling Language. XVII, 724 pages. 1999.

Vol. 1725: J. Pavelka, G. Tel, M. Bartošek (Eds.), SOFSEM'99: Theory and Practice of Informatics. Proceedings, 1999. XIII, 498 pages. 1999.

Vol. 1726: V. Varadharajan, Y. Mu (Eds.), Information and Communication Security. Proceedings, 1999. XI, 325 pages. 1999.

Vol. 1727: P.P. Chen, D.W. Embley, J. Kouloumdjian, S.W. Liddle, J.F. Roddick (Eds.), Advances in Conceptual Modeling. Proceedings, 1999. XI, 389 pages. 1999.

Vol. 1728: J. Akoka, M. Bouzeghoub, I. Comyn-Wattiau, E. Métais (Eds.), Conceptual Modeling – ER '99. Proceedings, 1999. XIV, 540 pages. 1999.

Vol. 1729: M. Mambo, Y. Zheng (Eds.), Information Security. Proceedings, 1999. IX, 277 pages. 1999.

Vol. 1730: M. Gelfond, N. Leone, G. Pfeifer (Eds.), Logic Programming and Nonmonotonic Reasoning. Proceedings, 1999. XI, 391 pages. 1999. (Subseries LNAI).

Vol. 1732: S. Matsuoka, R.R. Oldehoeft, M. Tholburn (Eds.), Computing in Object-Oriented Parallel Environments. Proceedings, 1999. VIII, 205 pages. 1999.

Vol. 1734: H. Hellwagner, A. Reinefeld (Eds.), SCI: Scalable Coherent Interface. XXI, 490 pages. 1999.

Vol. 1564: M. Vazirgiannis, Interactive Multimedia Documents. XIII, 161 pages. 1999.

Vol. 1591: D.J. Duke, I. Herman, M.S. Marshall, PREMO: A Framework for Multimedia Middleware. XII, 254 pages. 1999.

Vol. 1735: J.W. Amtrup, Incremental Speech Translation. XV, 200 pages. 1999. (Subseries LNAI).

Vol. 1736: L. Rizzo, S. Fdida (Eds.): Networked Group Communication. Proceedings, 1999. XIII, 339 pages. 1999.

Vol. 1738: C. Pandu Rangan, V. Raman, R. Ramanujam (Eds.), Foundations of Software Technology and Theoretical Computer Science. Proceedings, 1999. XII, 452 pages. 1999.

Vol. 1740: R. Baumgart (Ed.): Secure Networking – CQRE [Secure] '99. Proceedings, 1999. IX, 261 pages. 1999.

Vol. 1742: P.S. Thiagarajan, R. Yap (Eds.), Advances in Computing Science – ASIAN'99. Proceedings, 1999. XI, 397 pages. 1999.